Monetary Policy and Crude Oil

NEW DIRECTIONS IN POST-KEYNESIAN ECONOMICS

Series Editors: Louis-Philippe Rochon, *Laurentian University, Sudbury, Canada* and Sergio Rossi, *University of Fribourg, Switzerland*

Post-Keynesian economics is a school of thought inspired by the work of John Maynard Keynes, but also by Michal Kalecki, Joan Robinson, Nicholas Kaldor and other Cambridge economists, for whom money and effective demand are essential to explain economic activity. The aim of this series is to present original research work (single or co-authored volumes as well as edited books) that advances Post-Keynesian economics at both theoretical and policy-oriented levels.

Areas of research include, but are not limited to, monetary and financial economics, macro and microeconomics, international economics, development economics, economic policy, political economy, analyses of income distribution and financial crises, and the history of economic thought.

Titles in the series include:

Monetary Policy and Crude Oil

Prices, Production and Consumption

Basil Oberholzer

Global Infrastructure Basel Foundation, Switzerland

NEW DIRECTIONS IN POST-KEYNESIAN ECONOMICS

Edward Elgar
PUBLISHING

Cheltenham, UK • Northampton, MA, USA

Published by
Edward Elgar Publishing Limited
The Lypiatts
15 Lansdown Road
Cheltenham
Glos GL50 2JA
UK

Edward Elgar Publishing, Inc.
William Pratt House
9 Dewey Court
Northampton
Massachusetts 01060
USA

A catalogue record for this book
is available from the British Library

Library of Congress Control Number: 2017933393

This book is available electronically in the **Elgar**online
Economics subject collection
DOI 10.4337/9781786437891

ISBN 978 1 78643 788 4 (cased)
ISBN 978 1 78643 789 1 (eBook)

Typeset by Servis Filmsetting Ltd, Stockport, Cheshire
Printed and bound in Great Britain by TJ International Ltd, Padstow

Contents

Tables

Acknowledgements

I am deeply indebted to Prof. Dr Sergio Rossi for providing me with advice throughout this work. Without his tremendous help in reading and commenting as well as stimulating my thoughts, this book would not have been realized. In particular, it was the insight into his research fields that opened up new horizons for me, provided me with many ideas, and crucially supported the successful development of this work.

Likewise, I would like to thank Prof. Dr Philippe Gugler and Prof. Dr Laurent Donzé for reading and commenting on my work. Moreover, I am grateful to Edwin Le Heron, Associate Professor of Economics at the Bordeaux Institute of Political Studies for his comments on the SFC model. Finally, I want to thank all people unmentioned who inspired my notions by raising important issues in various discussions.

Abbreviations

ADF test	Augmented Dickey–Fuller test
AIC	Akaike information criterion
BIS	Bank for International Settlements
BP	British Petroleum
BTU	British thermal unit
CEA	Commodity Exchange Act
CFMA	Commodity Futures Modernization Act
CFTC	Commodity Futures Trading Commission
CMA	Calendar Monthly Average
ECB	European Control Bank
EIA	Energy Information Administration
Fed	Federal Reserve System
FOMC	Federal Open Market Committee
FSB	Financial Stability Board
GDP	Gross domestic product
HWWI	Hamburgisches WeltWirtschaftsInstitut/Hamburg Institute of International Economics
ICAPM	Intertemporal Capital Asset Pricing Model
IEA	International Energy Agency
ITF	Interagency Task Force on Commodity Markets
MA	Moving average
NYMEX	New York Mercantile Exchange
OECD	Organisation for Economic Co-operation and Development
OPEC	Organization of Petroleum Exporting Countries
OTC	Over the counter
SFC model	Stock-flow consistent model
SIC	Schwarz information criterion
SOMA	System Open Market Account
SVAR	Structural vector autoregression
VAR	Vector autoregression
WTI	Western Texas Intermediate

Introduction

Soaring gas prices have turned the steady migration by Americans to smaller cars into a stampede. [. . .] 'The era of the truck-based large S.U.V.'s is over,' said Michael Jackson, chief executive of AutoNation, the nation's largest auto retailer. [. . .] there are some indications that the trend toward smaller vehicles will reduce the nation's fuel use.

(Vlasic, 2 May 2008, New York Times)

Americans are buying more new cars than ever before. [. . .] As gas prices fell, Americans upsized. This fall, small SUVs became the largest segment of the market, at 14 percent, beating out small and midsize cars.

(The Associated Press, 5 January 2016)

To date, the twenty-first century has been marked by large fluctuations in global commodity prices. Among different commodities, crude oil plays a dominating role and represents, to a certain extent, an indicator of the overall development. It is not presumptuous to consider the oil price hike in 2008 and its preceding and subsequent strong variations as one of the outstanding global economic phenomena aside from the financial crisis, itself likewise erupting in 2008. At the moment of completion of this book, crude oil is still the most important energy source in the world. It has a benchmark function with respect to other fuels, notably natural gas and coal. On the one hand, oil contributes to prosperity and powers industrial production in the truest sense of the word. On the other hand, carbon emissions drive climate warming and hence are a long-run threat to the world's and people's well-being. Thus, the two citations above represent only a small sketch of the far-reaching impacts of oil market developments.

Against this background, the driving forces of the crude oil market become an issue of great interest. It is undoubted that crude oil shares many features with a conventional goods market with conventional feedback mechanisms between supply, demand and price. Yet, it would be too shortsighted to stop the analysis there. Our specific interest here is in the impact that monetary policy, that is, US monetary policy in our case, has on the crude oil market. The investigation of monetary policy effects requires a clear conception of the role of money in the economy. If it is neutral, as neoclassical economists argue it to be, at least in the middle to long run, changes in monetary conditions do not produce lasting effects.

However, if money is not neutral but, instead, allowed to have lasting impacts on both the supply and demand sides of the economy, an investigation of the connection between monetary policy and the crude oil market is not trivial anymore.

Crude oil has a dual character: it is both a physical commodity and a financial asset. Its first nature makes it resemble a conventional goods market. The second one is due to trading of futures contracts in commodity exchanges. As a consequence, monetary policy has the potential to affect the oil market through both aspects, that is, once through fundamentals in the spot market and once through 'paper oil' in the form of a financial asset. Hence, the same thing is traded in two different markets, which are nevertheless closely connected. Moreover, complex interdependencies between both mechanisms of transmission may arise. The spot and the futures markets compose the crude oil market as a whole.

In connection to the issue of money, the understanding of economic processes as such is crucial. In neoclassical theory, economic activity is embedded in a general equilibrium framework where the result is determined by utility and profit maximization. It is founded by microeconomics by means of aggregating individual behaviour linearly to the macro level. However, one may doubt the usefulness of the aggregation procedure by simply summing up all individual magnitudes. The total of individual actions, including complex interactions, may give rise to unexpected and sometimes paradox outcomes at the macroeconomic level. The aggregation problem is subject to uncertainty (Keynes, 1936/1997, pp. 161–162). Uncertainty is the idea that makes economic activity a radically indeterminate issue. In such an environment of uncertainty where money is allowed to exert lasting effects, the crude oil market may be influenced by monetary policy through various ways. Specifically, speculation is allowed to become a crucial feature and a kind of connection between monetary policy and the oil market by impacting on the oil price.

To outline briefly, we argue that expansive monetary policy leads to speculation in the futures market. The resulting higher oil price triggers overinvestment in the crude oil spot market, pulling down the price eventually to a lower level than the initial one. There are two main problems arising. First, the dual nature of crude oil raises economic and financial instability in the crude oil spot and futures markets as well as, to some extent, in the rest of the economy. Second, overinvestment raises oil supply. The lower oil price has a positive influence on consumption, which amounts to a threat of ecological sustainability in general and climate in particular. It is these challenges that an economic policy design must address. More concretely, stable financial and economic conditions in the crude oil market must be achieved, conditions that take the ecological dimension into account and provide a path towards sustainability. We will see that we can make use of the preceding insights into the workings of money, monetary policy, the crude oil market and the relationships in between. In this regard, our policy proposition is unconventional. Instead of aiming at eliminating any harmful effects of financial markets (the crude oil futures market in our case) on the real economy, we suggest how financial market mechanisms may be used to achieve a better economic outcome.

The analysis is divided in three parts. Part I contains the detailed theoretical analysis of our issue and emphasizes the crude oil market as well as a background from monetary theory to the extent that is required for our analysis. Chapter 1 starts with embedding the issue of crude oil in the currently existing environment and literature. Besides some stylized facts presented here, there are three principal domains concerning the crude oil market on which academic literature is focused. One is the debate about crude oil as an exhaustible resource and a fossil fuel. Another, an issue of great interest in economic research, is the existence and effectiveness of speculation in the oil market. Opinions are still widely diverging in this regard, while the view that, to a certain extent, speculation has tended to have a significant impact on the oil price, has gained ground in recent years. As a third topic, the role of the Organization of Petroleum Exporting Countries (OPEC) is debated.

Chapter 2 begins by emphasizing the nature and role of money in order to derive the functioning of monetary policy from it. Monetary theory can be broadly separated in a perspective that considers money as exogenous, that is, as a kind of commodity used for the exchange with goods and the quantity of which is controlled by the central bank. In contrast to this mainstream view inherent to neoclassical economics, the theory of endogenous money analyzes money as being created *ex nihilo* in the process of credit granting. Since money is demand-determined, the monetary authority cannot control its volume. This difference gives rise to a different understanding of monetary policy and its impacts. In the same way, the understanding of financial markets differs depending on the conception of money. Neoclassical economists rule out bubbles and endogenous distortions in financial markets by assuming the efficient market hypothesis. The existence of money does not affect the relationship between economic fundamentals and financial markets. By contrast, endogenous money, coupled with uncertainty, allows for a financial market evolution that is independent from the real economy to some extent. Speculation may become effective.

With the necessary insights into monetary theory, the oil market then is analyzed in light of the dual nature of crude oil. Monetary policy enters the stage by affecting the crude oil market both through fundamentals and the futures market. The analysis is extended by investigating the transmission channels of monetary policy in detail with regard to both fundamentals and financial market aspects. Out of these effects, numerous mutual impacts between the spot and the futures market emerge. Yet, we can conclude unambiguously by theoretical analysis that monetary policy has an effect on the oil price as well as oil quantities, to wit, production and consumption.

Part II is engaged in putting the theoretical analysis into the context of the actually existing policy and market structures and examining the issue empirically. In Chapter 3, the practice of US monetary policy is presented first. The period covered in our empirical investigation, that is, in general, from 2000 until 2014, is marked by a radical change of how monetary policy is conducted. Conventional policy, by manipulating the federal funds rate, was the rule prior to 2008. In the course of the crisis, the target rate reached the zero lower bound so that 'unconventional' policy, mainly in the form of asset purchases, was adopted.

The chapter investigates then the global crude oil pricing system. Contrary to the widely held belief that a market price is a definite numerical value resulting from exchange, there is in fact no single price because every deal between two parties yields its own price. This reveals that despite many other imperfections in the real world, even the assumption of a single price is a simplification of reality. There are numerous influences that impede the realization of market efficiency. Nonetheless, crude oil market data show that the market is integrated with regard to the geographical and the temporal dimension: Prices of different types of crude oil around the world are almost perfectly correlated and so are spot and futures prices of different maturities. It is thus fair to talk about a globally integrated crude oil market. Integration goes even further to some degree as natural gas and coal seem to follow a remarkably similar price pattern, like crude oil, in the middle to long run.

The impact of monetary policy on the crude oil market is analyzed econometrically in Chapter 4. To summarize, clarify and represent the results of the theoretical analysis as well as for the purpose of the isolated effects that are going to be estimated, we construct a stock-flow consistent (SFC) model of monetary policy and the crude oil market. This model reveals the main effects and supports intuition for the remainder of this examination. All in all, empirical estimates suffer simultaneity problems that are especially obvious in the context of fast evolving financial markets on the one hand and slowly reacting fundamentals on the other hand. Moreover, we argue that speculation is too complex a phenomenon in order to be represented by a single variable. Monetary policy, as a third important inconvenience, is difficult to be represented by a variable that is not anticipated by agents. To support empirical evidence, we consider the role of inventories as an approximation for speculative activity.

Two problems with monetary policy and the crude oil market arise from the preceding analysis: economic and financial instability; and a higher oil intensity of the economy implying a threat for the natural environment. In Part III, we address political answers for these problems. Chapter 5 debates existing policy propositions that are already partially realized in some cases. Some approaches address only the stability issue while others take only the ecological problem into account. In particular, we discuss futures market regulation and the use of the US strategic petroleum reserve in order to ensure price stability. With respect to environmental policy, a carbon emission trading system and an energy tax are frequently proposed in policy debates. We apply both ideas to the crude oil market and assess the resulting implications. All these propositions are successful in achieving the policy goals partially. However, they share shortcomings as they are not sufficient to guarantee stability and sustainability, and sometimes may even give rise to new problems.

For these reasons, a new approach is taken in Chapter 6. It aims at bringing together the advantages of each of the existing policy propositions while avoiding their drawbacks. Specifically, it must be an approach that is able to establish economic and financial stability as well as ecological sustainability without creating new macroeconomic problems. The idea we present in this chapter is unconventional. It does not try to eliminate financial market disturbances. Rather, a design of coordinated monetary and fiscal policy makes use of the existence of futures markets to lead the crude oil market and, in some measure, the economy as a whole to a stable and sustainable environment. We call it the 'oil price targeting system'.

PART I

Facts and theory of monetary policy and crude oil

1. The crude oil market and its driving forces

The global crude oil market has some specific features that are often debated and of which some are of primary and some of secondary importance. In order to understand the oil market throughout this work, we briefly consider these issues, review the corresponding literature and locate them with respect to our specific research goal.

1.1 THE IMPORTANCE OF CRUDE OIL: SOME FACTS

The global market for crude oil can be characterized by several features worth mentioning. They concern production and consumption patterns, price development and the importance of oil as an energy source. We briefly refer to each of them in turn. Oil production with data from the Energy Information Administration (EIA) is depicted in panel (a) of Figure 1.1. Obviously, worldwide crude oil output has featured a more or less steady long-run increase since 2000. While OPEC production features several fluctuations, production of the rest of the world follows a rather stable path. At the end of 2014, OPEC accounted for about 38 per cent of total world oil production. Panel (b) exhibits the demand side, that is, crude oil or petroleum consumption. Global consumption has as well continuously risen since 2000. Organisation for Economic Co-operation and Development (OECD) countries consumed slightly less oil in 2014 than they did in 2000. Conversely, non-OECD countries faced a strong and constant increase during the same time span and even outpaced OECD consumption in 2014.

Daily crude oil spot prices of type West Texas Intermediate (WTI) are plotted in Figure 1.2. In 2002, the price started to rise continuously until mid-2008, aside from an interrupt in 2006. The sharp price drop after the price peak in 2008 coincides with the outbreak of the global financial crisis. However, at the beginning of 2009, the price started rising again and stayed around 100 dollars. In the second half of 2014, it sharply decreased and has oscillated around 40 dollars since then. Even though our data window

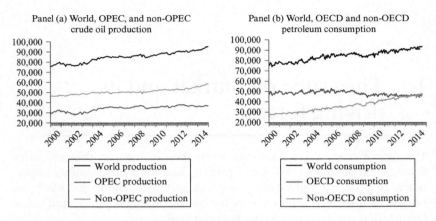

Source: Energy Information Administration (2015b). International Energy Statistics.

Figure 1.1 Global oil production and consumption (in thousand barrels per day), 2000–14

Source: Energy Information Administration (2015a). Petroleum and other Liquids.

Figure 1.2 WTI crude oil spot price per barrel in US dollars, 2000–14

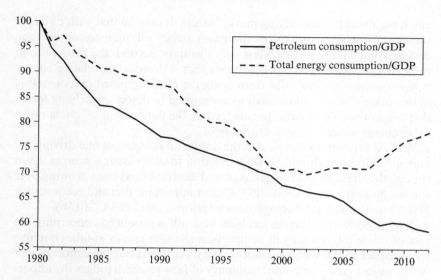

Sources: Energy Information Administration (2015b). International Energy Statistics; World Bank (2015). World DataBank.

Figure 1.3   Global petroleum and total energy intensity of output, 1980–2012 (1980 = 100)

already ends in 2014, the last phase of the price decline will be of great analytical interest. At the moment of completion of this book, the price persists in being low.

Relating oil consumption to total output, we get a measure of the oil intensity of the economy. Considering a longer time horizon since the 1980s, it can clearly be seen from Figure 1.3 that the oil intensity of output has remarkably decreased since. The pattern of total energy intensity of output, to wit, the intensity after including natural gas, coal and electricity, in addition to oil, is much less clear. It features a declining path from 1980 until 2000. From then on, intensity stagnated first and has again risen since. Without investigating this trend further, there are two remarks to make. First, like the graphs with oil production and consumption, the indexes of oil and total energy intensity only identify the crossing point of the supply and demand sides of the oil and total energy market, respectively (once we consider the denominator of the ratio, to wit, GDP, as given). The long-run decline of oil intensity neither says that the demand side of the oil market has become weaker relative to total output evolution, nor that the supply side has faced fixed constraints holding oil consumption down. Since data only reveal the final market result, we cannot say anything about the underlying market forces. It may be that with different stances of energy, economic or monetary policy oil intensity would have declined much more or much less over this time. Second, the recent rise in total energy intensity of output shows that a decrease in relative energy consumption over time is far from being the predetermined outcome. It is in this place that economic analysis is required to detect underlying forces that are not visible in data. In this regard, the connection of crude oil to other energy sources plays a crucial role.

While it is generally not denied that crude oil is a central fuel driving the global economy, it should be set in relation to other energy sources. Even though the share of natural gas, coal and electricity has been growing over the last 50 years, crude oil is still the most important fuel and accounts for about one-third of total energy consumption in 2012 (EIA, 2015b).

A large body of literature has been and still is produced concerning the role of crude oil, not only in comparison to other energy sources but also with respect to the whole macroeconomy. As this is not the centre of our work, we just give a very short summary of how research judges the importance of crude oil in the economy. In general, a rising oil price is associated with distorting effects on economic production of oil-importing countries. It is argued that GDP declines while inflation tends to increase (see, for instance, Sill, 2007). Yet, this is too absolute a conclusion as the combination of missing economic growth and high inflation was observed in the 1970s but not since. As Kilian (2010b, p. 14) remarks, an isolated recessionary oil price shock is expected to lead to deflation rather than inflation.

High oil prices can affect the economy on both the supply and the demand side as summed up by Kilian (2010b, pp. 5–10). A higher oil price raises input costs for producers and as such represents a supply-side constraint. On the demand side, households face a tighter budget constraint after having paid the bill for energy consumption such that there is less money left to spend for other goods. An additional and specific channel is argued to lie in monetary policy. Thereby, recessionary effects are not directly produced by high oil prices but rather by contractionary monetary policy aimed at stopping inflationary pressures that are themselves due to the change in oil prices (see, for example, Bernanke et al., 1997). Yet this argument is criticized to hold only if monetary policy is not anticipated (see Carlstrom & Fuerst, 2005). Otherwise, accommodating actions by the central bank are suggested to bring about the same results.

Most researchers find that the effects of oil price shocks on the economy have decreased in the 2000s compared to the 1970s (see Blanchard & Galí, 2007; Kilian, 2010a).[1] However, the explanations for this observation differ. Blanchard and Galí (2007) consider a high oil price as an exogenous shock. According to their analysis, first, the shocks went along with different additional effects in the 1970s than in the 2000s. Second, unsurprisingly in face of their neoclassical or new Keynesian background, reduced wage rigidities have lowered the effect on output and inflation. Third, monetary policy has become more credible and thus more able to keep inflation expectations low. Fourth, the share of oil in production, that is, oil intensity, has decreased, which limits potential effects of an oil shock *a priori*. Segal (2011) argues that the pass-through of the oil price decreased over past decades, which lowered the need of contractionary monetary policy and hence reduced harmful effects on the economy.

Kilian (2010a, 2010b) strongly criticizes the assumption of exogenous oil price shocks. Their effect on the economy and the way the monetary authority should respond to them depend on the cause of the oil price change, which is itself an endogenous variable. If it is a supply disruption due to, say, a war in an oil-producing country, the high oil price effectively tends to affect economic performance negatively to a certain extent. In this case, the central bank should loosen monetary conditions despite potential inflationary pressure in order to counteract a recession. Otherwise, it would just deepen it. On the other hand, if it is a high oil demand at home or abroad leading to a higher oil price, there is no recession to expect since the oil price is itself just a symptom of a boom period. It is in this case that inflation should be treated by contractionary monetary policy. The recession follows once the demand boom decelerates (ibid., 2010a, p. 81). The need for different monetary policy responses to different kinds of oil price shocks is confirmed, for example, by Bodenstein et al. (2012). The decreasing effect of high oil prices therefore is to ascribe to their different origins.

Even though literature finds a decreasing impact of oil price changes on production, it does not mean that the relevance of crude oil is declining as well. Besides its impact on GDP growth rates, its character as a nonrenewable fuel and the fact that fossil energy sources still account for more than 80 per cent of total energy consumption make it highly important to analyze the market for crude oil, specifically against the background of climate change.

1.2 CRUDE OIL AS AN EXHAUSTIBLE AND FOSSIL RESOURCE

The first question asked in common discussions about oil is the one regarding how much of it is left in the world. Crude oil is exhaustible. This is the characteristic standing in the spotlight of research on, and political debates about, climate change and resource security (see, for instance, International Energy Agency (IEA), 2012, pp. 97–101). Oil – together with other non-renewable energy sources like natural gas and coal – is the central fuel driving global production. The limitation of oil reserves in the underground and the numerous threats of climate change will require humanity to search for new sources. A large part of debate considers the basic task as a technological one requiring the replacement of fossil fuels by renewables (see Hamilton, 2012). Other voices go much further by suggesting that the need for more energy efficiency and saving in energy consumption will require fundamental changes in social and economic organization. According to these arguments, the difficulty is not primarily of a technological nature the resolving of which would allow business as usual. As the best-known among other institutions and individual scientists, the Club of Rome identifies permanent growth in economic output as the most important cause of rising energy consumption and worsening environmental damages (Meadows et al., 1972, pp. 38–44, 54–87). In its original and famous book *The Limits to Growth* (Meadows et al., 1972), it predicts the collapse of the existing world economy and the current way of living in developed countries if no change in the organization of human living takes place. The reason for this sinister perspective is the contradiction between economic growth and natural limits given by ending resources. One day, the world will reach peak oil, that is, the maximum possible crude oil production. From that point onwards, it cannot but follow a steady decline sooner or later owing to diminishing oil reserves. Once the last drop of oil is burnt, the economy breaks down.

This prediction has not become reality up to now and we do not know the time horizon of the further development. Even though such analyses address problems of greatest seriousness, which should not be ignored even though the collapse has not yet become reality, the basis of the Club of Rome's and others' predictions is not able to overcome some complex economic issues. Predicting linearly or exponentially increasing energy consumption ignores price effects. If fossil fuels become scarcer, their price is likely to rise which leads to new factor allocation. For example, alternative energy production replaces crude oil partially. This has a declining effect on the oil price. As another example, higher expenditures for research and development in the field of renewable energy and energy efficiency may make alternatives to fossil fuel less expensive and hence more competitive. Hence, whether the end of fossil resources ever takes place and, if this should be the case, whether the passing over to a new way of energy production takes the form of chaos or collapse, is highly uncertain.

Despite uncertainty, such reflections are important, especially from a normative point of view. In positive economic analysis, the role of resource limitation is somewhat different. It can be seen as an outermost constraint within which the economics of the crude oil market take place. Once this constraint becomes binding, we face a situation where varying demand meets constrained supply. In the case of demand growth, this leads to a – potentially strongly – rising oil price. It is this case to which the well-known 'Hotelling rule' of the optimal consumption rate of exhaustible resources is applied (Hotelling, 1931). The classical economists have already regarded price-building in markets for limited goods. This has led to the definition of rent, to wit, a form of extra profit earned by the owner of such a resource, which used to be land or metals in the eighteenth and nineteenth centuries (see, for instance, Marx, 1894/2004, pp. 602–788; Ricardo, 1821/1923, pp. 52–75; Smith, 1776/1976, pp. 160–275). In post-Keynesian literature, in line with classical economists, limited natural resources are also distinguished from other, unconstrained, goods. While the latter are argued to face a horizontal long-run supply curve, the former behave differently, since rising demand is not necessarily satisfied by equally increasing supply at a stable price (Kalecki, 1987, p. 100). Importantly to note, the supply constraint cannot automatically be set equal to all proven reserves in the world. Not all reserves are profitable at all oil price levels. Those reserves that are difficult to access face a higher production cost and thus are only extracted if a sufficiently high oil price guarantees profits to producers. A rising price may give an incentive to intensify exploration activities. This shifts the constraint outwards even though it might have been considered as fixed before. It is therefore not possible *a priori* to determine the definite total volume of oil reserves that is relevant for our economic investigation. Furthermore, it may happen that the constraint of exhaustion becomes partially binding. Oil might still be geologically available in abundance but it is only accessible if the oil price reaches a specific minimum level. The resource then is partially exhausted in the sense that it no longer exists at the hitherto low prices. This gives the supply curve again another form.

As long as demand can be satisfied by oil production without requiring a higher price, the limitation of total existing reserves is not relevant for market participants at that moment, that is, the outermost constraint of reserves is not binding. Hence, within this constraint, there are other limits that are relevant in the shorter term already before oil reserves run the fear of exhaustion. These are production capacities that can be fully utilized in a given situation such that higher demand raises the price. Additionally, oil companies and households may possess inventories that they use as a (personal) reserve in order to hedge against price changes. Once all inventories are sold or consumed they cannot serve as a buffer against price fluctuations anymore. Another reason for rising prices may lie in market power on the producer side. We can therefore say that, as long as oil reserves suffice to satisfy existing demand, crude oil shares the same features as any other product market.

For these reasons, price changes should not precipitately be attributed to exhausting oil. Even though this is relevant without doubt in the long run, it is not necessarily true that a high oil price in a given situation is due to the absolute exhaustion of oil reserves. A case of absolute exhaustion occurs if the oil price keeps rising without there being a reaction on the supply side. Partial exhaustion is much more difficult to assess.

Concerning the present-day degree of absolute exhaustion of oil reserves, opinions differ widely. The IEA (2013, pp. 3–4) forecasts a declining production of conventional oil requiring the extraction of unconventional sources on the one hand, and a high oil price in 2035 on the other, to fill the gap between supply and demand. In contrast, for example, the Statistical Review of World Energy of British Petroleum (BP) (2015, p. 7) estimates that the ratio of global total oil reserves to annual global production has not only been constant but even slightly rising since the middle of the 1980s. In 2014, this ratio states that proven reserves are sufficient for 52.5 years of global production. Being aware of the wide area within which the technical and economic debate about long-term prospects takes place, we will have to find out how far they are in relevance when analyzing the connections between monetary policy and the global market for crude oil.

1.3   THE ISSUE OF SPECULATION

Let us now turn to the central debate where monetary policy comes into play as will be discussed in abundance in this work. This concerns the driving forces of the crude oil market. It may be fair to say that the largest amount of literature concerned with this topic concentrates exclusively on the oil price and ignores the effects that the price-driving forces have on oil quantities. Thereby, it centres on the question whether the oil price is only determined by supply of, and demand for, physical oil or whether speculation has a significant impact, too. The important point lies in our interest of how monetary policy affects the crude oil market through financial markets.

First of all, it is not easy to get a definite meaning of speculation. This may be one reason why a considerable part of research omits it. Yet, another one might be that the existence of speculation is not equally acknowledged by all economists. It will be seen in more detail in the next chapter that neoclassical economics in its proper sense does not leave room for speculation. Or, to put it at a little more length, it may be allowed to exist but it does not have significant effects on other variables. However, being convinced or not of the existence of speculation, it needs to be defined so that it can be tested. Kaldor (1939, p. 1) provides a definition in an influential paper:

> Speculation [. . .] may be defined as the purchase (or sale) of goods with a view to re-sale (re-purchase) at a later date, where the motive behind such action is the expectation of a change in the relevant prices relatively to the ruling price and not a gain accruing through their use, or any kind of transformation effected in them or their transfer between different markets. [. . .] What distinguishes speculative purchases and sales from other kinds of purchases and sales is the expectation of an impending change in ruling the market price as the sole motive of action.

As another, contemporary source, Kilian and Murphy (2014, p. 455) determine speculation, in the specific case of crude oil, by 'treat[ing] anyone buying crude oil not for current consumption but for future use as a speculator from an economic point of view. Speculative purchases of oil usually occur because the buyer is anticipating rising oil prices'. These definitions should basically not be too controversial even though the authors mentioned belong to different schools of economic thought.

Yet, opinions differ with regard to the impact that speculation has on prices, quantities, economic activity, employment and economic stability. The issue becomes controversial at this point because it moves from the definition of speculation to its embedment into the economy. This act may still be positive but it is close to the somewhat normative judgement of whether speculation is beneficial to the economy and society or not. Kaldor (1939, pp. 2, 10) argues that speculation can be both price-stabilizing and price-destabilizing, depending on the magnitude of speculative activity and the range within which the price of the asset is moving alongside speculative influences. He outlines several conditions to be fulfilled in order for an asset to be traded speculatively. It must be fully standardized, durable, valuable in proportion to its bulk and it must be an article of 'general demand', that is, it must be an important good in the economy (ibid., p. 3). If so, then it becomes possible to bet on changing prices by accumulating or reducing the stocks of the relevant asset. The more speculators build stocks when they expect a higher price, the more speculation eventually affects the actual price (ibid., p. 7). Additionally, the more speculators change their price expectations in the face of a change in the current price, the more they raise or lower their stocks (ibid., pp. 8–9). By means of these two elasticities, the influence of speculation is assessed.

This theory is contradicted by other, mainly neoclassical, economists. Defending the 'efficient market hypothesis', yet to be discussed in more detail, they deny a significant impact of speculation on other variables. It is the efficient market as a whole that determines prices. Speculators who swim against the storm make a losing deal for sure, that is, they cannot beat the market (see Malkiel, 2003, p. 77). If anything, speculation is argued to be beneficial because speculative investment provides the liquidity necessary to allow not only faster price discovery but even guarantees the functioning of financial markets (Fattouh et al., 2012, p. 4). Financial stability is therefore enhanced.

One may think that it is more appropriate to take newer conceptions of speculation than those of the first half of the twentieth century. However, as just seen, we cannot say that new research results have definitely abandoned older explanations. Even though Kaldor's (1939) theory is rather mechanical in some aspects and without ruling out other approaches here, we can keep it in mind for the remainder. It does not insist on the inevitable and permanent presence of speculation nor does it eliminate it *a priori* by theoretical assumptions. It just draws the mechanisms through which it may become effective. By testing for the existence and effectiveness of speculation, most authors more or less follow Kaldor's line of thought independent of whether they consider it as realistic or not.

Yet, in fact, assessing speculation empirically is quite difficult. In the case of producing companies, for example, it is seemingly clear that any intertemporal considerations or any exposure in financial contracts serve the purpose of hedging against future price fluctuations. Concerning crude oil, a corporation may accumulate inventories in order to smooth price hikes when sudden supply interruptions or demand growth take place. But what is the appropriate level of stocks in such a situation? The corporation's stock-building may as well have the effect of keeping the oil price higher than necessary. The line between hedging and speculating thus may often be difficult to draw. Another aspect of Kaldor's theory that is not fulfilled nowadays anymore is the argument that speculation is done by accumulating stocks. That is not wrong. We will as well argue that a high price tends to go along with higher inventories if it is to be driven up by speculation. Yet, as will become clear throughout our theoretical analysis, stocks do not necessarily have to be the cause of the speculative price change. They may as well and even more likely be the result of speculation. In our highly financialized economies, oil futures contracts are the object of interest for most speculators. Thereby, they are completely disconnected from the physical aspects of the underlying real asset. Hence, their owners are not at all concerned with the accumulation of stocks.

In contrast to other commodities, crude oil as a fossil fuel requires another specification. Kaldor (1939, pp. 10–11) argues that commodity prices are determined by 'supply price' in the long run, that is, by production cost and a certain profit share. To the degree that this supply price is known, the actual price should sooner or later come back to this level once it deviates from it. Yet, crude oil is not infinite and its possible exhaustion in the future may lead speculators to drive their price expectations upwards so that – if speculation is effective – the actual oil price is as well higher than it would be otherwise. Such expectations might be justified or not. But they have the potential to be relevant in both cases. Moreover, exhaustion does not allow for acting on the assumption of a constant supply price. The limitation of oil sources is thus an issue that makes speculation even more difficult to assess.

Before we start the detailed analysis of monetary policy, speculation and its impact on the market for crude oil, let us get an overview of the existing literature on these issues as a preliminary. Indeed, this literature has the same (or similar) starting points as our analysis. Some contributions emphasize monetary policy explicitly. The outstanding feature of the past years that turned its special interest to the crude oil market, is the price peak in 2008. Even though research on this single event may appear as quite specific, it contains nevertheless many features that help understand the market in general. In the course of our analysis, we will have a closer look at many of the following contributions.

The view that the crude oil market is merely driven by fundamental forces, to wit, supply by oil producers and demand by oil consumers, is briefly explained by Fattouh (2010, p. 14), who calls this rather simple approach the 'conventional framework': changes in the crude oil price trigger feedbacks from the supply and demand sides. The resulting oil price depends on price elasticities of oil supply and demand. The lower they are, the higher the oil price can rise or fall without being counteracted by supply and demand responses (ibid., p. 16). Among those contributors who deny any significant speculative influence on the oil price during the price hike of 2008, opinions differ again with regard to fundamentals. Some see the major cause of the price increase on the supply side, others on the demand side. One of the most influential articles published in the line of the fundamentals view is Kilian (2009b). By constructing a vector autoregression (VAR) model that allows for endogenizing the oil price, he finds that the price peak of 2008 is almost entirely explained by increasing demand.

In the context of demand-side considerations, the oil price peak is often ascribed to the fast-rising need for energy of emerging economies like those of China or India. In another paper, Kilian (2009a) tests for the impact of these two countries by taking GDP forecasts as the basis of the approach. The study finds that the two emerging economies are an important but by far not the only source of the price increase since forecast error corrections of OECD countries also explain a comparable fraction of the price. Mu and Ye (2011) sharply contradict the view of emerging economies' demand growth as a significant source of the relevant oil price increase. They focus on China as the most important of these countries. By means of a VAR, they provide evidence that an increase in China's oil imports does not have a significant effect on the real oil price.

The stance that the high oil price was caused by supply-side shortages is taken up by Kaufmann (2011) who strongly criticizes Kilian's (2009b) approach. Kaufmann (2011, pp. 106–108) shows that, Kilian's demand variable is influenced by the oil price rather than the other way around. With respect to the supply side, he divides oil production into OPEC and non-OPEC output. While non-OPEC output was limited, OPEC oil production played the role of the marginal supplier which pushed up the oil price due to strategic behaviour (ibid., p. 108). Moreover, total capacity utilization increasingly approached the upper bound.

A growing body of literature considers the pure fundamentals view as insufficient. Another section denies any speculative influence but changes the approach in order to have a theoretical background on the basis of which speculation can be tested empirically. A theoretical model of speculation with an undefined asset is provided by De Long et al. (1990). As a crucial feature, they distinguish three types of investors and hence differ from basic neoclassical assumptions of homogenous agents or a single representative agent. There are rational informed investors who build expectations about the true value of the asset, which is postulated to be well defined. Passive investors buy when the price is low and sell when it is high. Feedback traders consider the price history of the preceding periods, invest when the price was rising in the past, and sell otherwise. The authors show how the price then can deviate from the asset's fundamentals value, such that a speculative bubble emerges. Tokic (2011) extends the De Long et al. (1990) model and applies it to the crude oil market. Producers and consumers basically have a price-stabilizing effect as they regulate oil inventories in order to smooth fluctuations. Passive investors who are in search of portfolio diversification and inflation protection, however, do not consider oil market fundamentals but, rather, use inflation rates and stock market risk as indicators to decide on oil purchases. They might therefore move the oil price, which attracts feedback traders. When the price is increasing, those agents who act on the assumption that the price always reverts to its fundamental value cannot but participate in the speculative bubble. If they continued pursuing their strategy of selling oil, that is, going short, they would suffer growing financial losses. Once feedback traders realize that the price is above the fundamental value, they start selling and the bubble bursts (Tokic, 2011, pp. 2057–2058).

Fattouh and Mahadeva (2012) present a neoclassical model with speculators, producers and consumers, that extends over two periods. Owing to agents' awareness of this fact, it becomes harder for large price deviations from fundamentals to occur, since changes in period 1 are counteracted in period 2. The term structure of the oil spot price can be tilted but it cannot be shifted (ibid., p. 16). Basically, however, the model allows for financialization when speculators raise their oil market exposure. Calibration with real data leads the authors to the conclusion that speculative activity may of course exist but that it affects the oil price only marginally. In contrast to this, Cifarelli and Paladino (2010) apply an Intertemporal Capital Asset Pricing Model (ICAPM) to the crude oil market and include the possibility of feedback trading. Empirical tests confirm significant positive feedback trading, which has itself a significant impact on the oil price.

Another theoretical model is developed by Alquist and Kilian (2010): a higher risk of future supply shortfalls raises the incentive to hold additional inventories today. This raises the oil spot price in relation to the futures price. Even though one may allow speaking about speculation in this case, the authors prefer the term 'precautionary demand' indicating that inventory accumulation merely serves hedging needs (ibid., p. 540). A precautionary demand component econometrically determined in Kilian (2009b) is tested for correlation with the futures spot spread and exhibits remarkable results but also fails for a part of the period considered (Alquist & Kilian, 2010, p. 566).

Knittel and Pindyck (2013) construct a simple model of constant price elasticities of supply and demand, and test for the possibilities of speculation as well as its implied effects on the price of oil and inventories. They find that the actual pattern of the oil price can be replicated by fundamentals data so that there is no room for speculation.

A series of purely empirical contributions examines Granger causalities between the speculative activity and the crude oil price (see, for instance, Alquist & Gervais, 2011; Büyükşahin & Harris, 2011; Interagency Task Force on Commodity Markets (ITF), 2008). They usually employ net long futures positions of different categories of investors. It is assumed to be an appropriate measure of speculation. The studies get similar conclusions, namely, that changes in the crude oil price Granger cause changes in futures holdings of investors but that there is no significant Granger causality in the reverse direction. Stoll and Whaley (2010) investigate the impact of index investment on non-energy commodity futures prices and end up with an analogous conclusion.

Another study leads to opposite results. Tang and Xiong (2011) analyze the connection between different commodities with a focus on commodity indexes. Index investors are likely to behave according to Tokic (2011, p. 2056) insofar as they do not aim at exploiting the expected price change of a single asset but rather invest in commodities to diversify their portfolio and to protect against inflation. First, Tang and Xiong (2011) find that correlation between different commodity prices, for example, between crude oil and some selected non-energy commodities, increased over time and that the increase was more pronounced for index-traded commodities. These findings favour the view that commodity prices are influenced by speculation and thus may deviate from fundamentals. The latter differ with respect to each single commodity such that prices would differ as well, if fundamentals were the only determining variables.

A similar result is obtained by Büyükşahin and Robe (2011). They have access to disaggregated non-public data and test for co-movement of energy commodity and stock prices. The results suggest that correlation between the prices of the two markets increases with the presence of hedge funds that are active in both markets. If groups of financial investors are able to influence market performance of assets, then it is likely that these asset prices can deviate from fundamental values.

Lombardi and Van Robays (2011) use a VAR and introduce the oil futures price in addition to the spot price. The defined speculative shock and a precautionary demand variable are found to contribute to the spot price increase between 2000 and 2008 (pp. 25, 26).

In the same paper already mentioned above, Kaufmann (2011, pp. 109–114) tests for the presence of speculation in the oil price by testing for the law of one price. He observes that the prices of two different types of crude oil, that is, the WTI five-month forward contract and the spot price for Dubai-Fateh, exhibit a cointegrating relationship. This relationship is found to break down in the period of strong price growth in 2007 and 2008. With two alternative approaches, this breakdown is confirmed. Evidence of such explanation gaps is not a direct confirmation but at least a hint of speculation.

Fan and Xu (2011) test for structural breaks in order to find evidence of how the price-determining variables of the crude oil price might have changed in the course of the 2000s. The overall result states that futures position variables are significant during the oil price growth before 2008 while the fundamentals variable is insignificant. After the price sharply drops, the fundamentals variable becomes significant while the speculative variables no longer have any explanatory power.

Lammerding et al. (2013) construct a bubble state-space model. They assume the oil price switches between two regimes, where in one it follows its true fundamental value while in the other it departs from that value, meaning that a speculative bubble is accumulating. Application of the model to real data shows that high bubble regime probabilities indicate price bubbles in the course of the run-up to the price peak in 2008 and again in 2009 when the price starts rising again.

In literature where monetary policy is directly involved, speculation is often only implicitly discussed. A theoretical body to which debate often refers is provided by Frankel (1984, 2006, 2014) and Frankel and Rose (2010). It says that interest rates affect the oil price through a couple of channels both on the supply and the demand side of the crude oil market such that contractionary monetary policy lowers the oil price (see, for instance, Frankel, 2006, p. 5). These channels will be discussed in more detail. Some papers test the theoretical implications and find mixed evidence (see Anzuini et al., 2013; Arora & Tanner, 2013). An event-based study tests for the same effects of unconventional monetary policy on prices of commodity indexes between 2008 and 2010 and does not find significant results in favour of the underlying theory (Glick & Leduc, 2011). Other studies employ high-frequency data to investigate intraday responses of oil prices to monetary policy shocks (see Basistha & Kurov, 2015; Rosa, 2013). The motivation is that price responses within just a few minutes have a greater chance to be free of noise from other shocks. There are significant results suggesting that a negative interest rate shock raises oil prices. However, the results usually become insignificant when enlarging the framework to daily or monthly responses (Basistha & Kurov, 2015, pp. 95–102). This may either be due to effective insignificance or to econometric difficulties in isolating longer-term effects.

1.4 THE ROLE OF OPEC

As a last short discussion before starting our theoretical analysis, the role of the OPEC has to be enlightened so far as it is possible at this stage. The issue has lost importance in the most recent past but still appears in literature. Furthermore, OPEC strategies are subject to public policy debates (see, for example, Reed, 2014).

The organization was founded in 1960 and currently includes 12 countries: Algeria, Angola, Ecuador, Iran, Iraq, Kuwait, Libya, Nigeria, Qatar, Saudi Arabia, United Arab Emirates and Venezuela. This number was variable in the decades of OPEC's existence. According to the organization's own words, its mission is:

> to coordinate and unify the petroleum policies of its Member Countries and ensure the stabilization of oil markets in order to secure an efficient, economic and regular supply of petroleum to consumers, a steady income to producers and a fair return on capital for those investing in the petroleum industry. (OPEC, 2015)

In most publications outside of OPEC countries, the organization is simply denoted a 'cartel' (see, for instance, Griffin & Vielhaber, 1994) aimed at controlling the global crude oil market. Bandyopadhyay (2009, pp. 14–29) provides a helpful overview of a great volume of research contributions produced in the past four decades. The central questions of research are whether OPEC is really able to influence the oil price by means of cooperation between its member countries and the agreement on production quotas. Kaufmann et al. (2004) present evidence of Granger causality from OPEC utilization of production capacities and production quotas to the real oil price. Yet, they find no Granger causality in the opposite direction. On the other hand, Brémond et al. (2012) argue that OPEC has been acting as a price taker for most periods since 1973. Smith (2009, p. 152) differentiates by suggesting that OPEC has failed to agree on lower utilization of existing extraction capacities but has succeeded at limiting the building of capacities by constraining efforts to explore new reserves. According to Alhajji and Huettner (2000), it is not OPEC as a whole but rather Saudi Arabia as a single country that acts as a dominant producer.

There are numerous claims that OPEC's strength in affecting oil prices has overall decreased over the decades of its existence. For example, decreasing OPEC spare capacities, especially during the 1980s and again after 2002 may be seen as a sign of declining market power (see Bandyopadhyay, 2008, p. 20). However, in contrast, it may well be considered a demonstration of strength rather than weakness of the workings of the OPEC cartel in Smith's (2009) sense, that the organization also agrees on the building of new capacities.

All in all, the literature remains inconclusive. The difficulty might be found in the complexity to model real-world strategic behaviour of a cartel like OPEC. Strategic behaviour is normally discussed in a microeconomic context. In the case of OPEC, however, it can be traced back to human decisions that have a more or less direct impact at the macroeconomic level. Therefore, there are probably even less linear, calculable and repeating effects than one might hope to find in other economic problems. As Bandyopadhyay (2009, p. 13) puts it, 'it is rather unusual to expect a rigid behavioural conduct from OPEC consisting of members with divergent views and interests. Thus, it would be unrealistic to predict the behavioural nature of the OPEC by making use of a single economic mode'.

A common approach, also adopted in this work, is to assume non-OPEC oil-producing countries as price takers (see, for instance, Kaufmann, 2011, p. 108). This means that oil companies in these countries behave like competitive firms in a global market, that is, a real market with its various imperfections that we will discuss later. As the average of studies characterizes OPEC neither as a perfect price maker nor as a perfect price taker, we have to be aware of a still unresolved problem when analyzing the global crude oil market. It does not make investigation impossible. However, a potential influence of strategic behaviour on the crude oil price – whatever its extent may be – tends to complicate the analysis.

NOTE

1. Barsky and Kilian (2004) argue that even the recessionary effects of the oil shocks of the twentieth century are overestimated in economic debates.

2. Monetary policy and crude oil: a theoretical analysis

As a priority, a large section of this work will be dedicated to the theoretical analysis. It is especially in empirical papers that the underlying theoretical reasoning remains confused or relies on unquestioned assumptions. Since crude oil is analyzed in connection with monetary policy, the examination of the working of our contemporary monetary system and of financial markets is of distinguished importance. It becomes thereby clear that it is the financial aspect of crude oil that needs specific consideration. For these reasons, we start with the issue of money, monetary policy and financial markets in a general manner.

2.1 ON MONEY AND MONETARY POLICY

The analysis of the relationships between monetary policy and the crude oil market requires a foregoing investigation of some basic issues. It will be seen that the characteristics of money play a crucial role in understanding those processes that take place in economic reality. Different conceptions of money yield different conclusions. Throughout this section, we start in each case with neoclassical explanations as the dominating paradigm and confront them with criticism. Exogenous money is criticized by means of endogenous-money concepts. Endogenous money then is used to develop an alternative view of financial markets, which takes criticisms of neoclassical financial theory into account. For the understanding of the central issues of this section, it is sufficient to emphasize monetary policy in its conventional form. 'Unconventional' policy is introduced later.

2.1.1 Monetary Policy with Exogenous and Endogenous Money

This section discusses an elementary aspect of monetary theory so far as it is necessary for the further understanding of our analysis. It is about the nature, source and effects of money. Economists can be roughly divided into two factions, one of which interprets the origin of money to be exogenous. The other considers money to be endogenously determined in the course of economic activity.

Exogenous money

The conception of exogenous money is strongly associated with neoclassical economic theory. It claims that money is emitted by the central bank. While it had the character of a commodity money during the time of the gold standard, it is nowadays created out of nothing and hence called *fiat money* (see, for instance, Friedman, 1986; Ritter, 1995, pp. 134–135). The regulation of the stock of money is part of monetary policy. Thus, the meaning of exogeneity is that money is injected into the economy as if it had, so to speak, fallen from heaven (Davidson, 2006, p. 146).

Commercial banks use the money issued by the central bank (to wit, central bank money) as reserves, which allow them to give loans to the economy. Payments financed by these loans give rise to deposits in the same or other banks, which are thus in a position to grant new loans. By this process, the initial reserves are multiplied and give rise to the stock of money effectively observed in the economy. The so-called money multiplier may change from one specific situation to another but is assumed to be fairly stable over time, such that the central bank is able to control the total amount of money by managing central bank money (Friedman, 1959, p. 527). Money supply depends on a policy decision. The supply curve of money is therefore vertical with respect to the market rate of interest (see, for example, Blanchard & Illing, 2006, pp. 145–148).

This monetarist view, strongly influenced and spread by Milton Friedman, has the quantity theory of money at its centre (see, for instance, Friedman, 1956). It argues that under the assumption that the velocity of money circulation is stable, the money stock should grow proportionately to GDP. A rise in the amount of money which is in excess of the economic growth rate leads to a proportional increase of the general price level in the middle or long run. There is a unidirectional causality from the quantity of money to nominal output (Palley, 1993, p. 79). Consequently, all else equal with respect to money velocity and GDP, expansive monetary policy leads to a higher rate of inflation, which is equal to the growth rate of the money stock. Since real variables in the market are given at a certain moment of time, monetary policy can only influence monetary variables. A higher money supply raises demand for goods, which again results in higher prices. Meanwhile, the supply side of the economy is supposed to remain unchanged (Lavoie, 2006b). Inflation is exclusively a demand phenomenon (see Barro & Grossman, 1974; Brunner et al., 1973). To say it in often used words, there is no (long-run) trade-off between output and inflation that monetary policy could exploit (see Barro & Gordon, 1983, p. 590;

Bernanke & Mishkin, 1997, p. 104). A conduct of monetary policy that is either too contractionary or too expansive with regard to what real market forces require will necessarily lead to corresponding changes in the price level. Monetary policy can control nominal variables but not real variables (Friedman, 1968, p. 11). The only sustaining effect of monetary policy is a price effect. This position is represented in the Friedman aphorism that 'inflation is always and everywhere a monetary phenomenon', supported by a large number of neoclassical economists (see, for instance, Mishkin, 2007, p. 2).

Considering money as exogenous makes it similar to a commodity.[1] Its price rises or falls depending on its scarcity. When the central bank aims at stimulating investment by strongly raising money supply, the value of money drops and inflation increases. Increasing investment without accelerating inflation in the long run is only possible if savings have increased first. It is only higher savings that allow for higher investment. Savings and investment are themselves equilibrated by the real interest rate. However, daily monetary policy may influence the interest rate level such that it differs from the level corresponding to the equilibrium. Thus, there is one observable rate of interest, regulated by monetary policy, and one unobserved rate. The latter equilibrates savings and investment in real, that is, non-monetary, terms. It depends on the productivity growth and savings behaviour of economic agents. This unobserved interest rate is called the natural rate and is part of the monetary analysis of Wicksell (Rochon, 2004, p. 2). It needs to be mentioned that some approaches rely on the theory of the natural interest rate even though they have basically incorporated the endogenous-money view. The New Consensus approach to monetary policy is a prominent example of this.[2] It pursues monetary policy targets by setting the short-run interest rate rather than trying to control the money stock (see, for instance, Woodford, 2002, p. 86). The aim of policy conduct, however, is not to set the rate freely but to shadow the natural interest rate (Goodfriend, 2007, p. 29). This strategy is incorporated in the famous and widely adopted 'Taylor rule', where the interest rate policy follows the natural rate by correcting for deviations in the inflation rate from the target rate and actual output from potential output (Taylor, 1993, p. 202). With regard to the theoretical background and the practice of monetary policy, the New Consensus approach resembles, in many respects, the paradigm of treating money as exogenously given. In particular, monetarism has based itself on the Wicksellian concept of the natural interest rate since its earliest days (Friedman, 1968, pp. 7–8). It is therefore not necessary to treat it separately for the remainder of the analysis.

The natural interest rate is a reference point that should be obeyed at least in the long run. When the stock of money is too high as a result of expansive monetary policy, the observed nominal interest rate is below the natural rate. This will lead to a higher rate of inflation because the stock of money is in excess of what is needed for economic transactions at a given output and price level. The opposite happens when the nominal interest rate set by monetary policy is above the natural rate. The only sustainable solution in the long run is the equivalence of both rates. Monetary policy has to react in an equilibrating way (see, for example, Romer, 2000, p. 156).

It should be taken into account that the neoclassical paradigm has been developing since the 1960s when monetarism became influential. On the one hand, new Keynesian economics introduced imperfect competition and nominal rigidities, which made the influence of monetary policy stronger in the short run (see Ball et al., 1988). On the other hand, real business cycle models became popular. They consider the economy as a perfectly competitive system that responds to exogenous productivity shocks in such a way that equilibrium is re-established. Money is a simple addition. The combination of these developments, new Keynesianism and real business cycle theory, is sometimes referred to as the 'new neoclassical synthesis' (Goodfriend & King, 1997). Monetary policy is suggested to react to exogenous shocks by setting the interest rate such that the economy is led back to its equilibrium. Hence, monetary policy's task is to enable the economy to work as much as possible like a perfectly flexible system. It is therefore basically effective but only useful if it corrects harmful deviations from the equilibrium path. It does not have the potential to create a positive outcome by departing from an equilibrium point determined by nature (ibid., p. 280). Moreover, even this result is questioned by other authors. They argue that nominal rigidities only delay the neutrality of money but restore it as soon as prices have adjusted (see, for instance, McGrattan, 1997, p. 286). Monetary policy is therefore still influenced by monetarism, in the sense that it is mainly effective in controlling inflation (Woodford, 2009, pp. 272–273). Effects on real variables are constrained by general equilibrium with flexible prices. It is the optimal outcome that cannot be further improved. The wrong conduct of monetary policy is condemned to result in inefficient distortions. Fluctuations in output and unemployment are due to exogenous shocks to the real business cycle. They should not be counteracted by monetary policy (Goodfriend, 2007, p. 26). To conclude, these new evolutions in theory do not allow monetary policy to become more powerful. They still judge the creation of money as inflationary because of its exogenous nature.

Endogenous money
The endogenous-money view faces some crucial differences. It is part of the post-Keynesian economic theory and intensely debated within this strand

of literature but not limited to it. According to this theory, developed to a great extent by Kaldor (1970, 1982) and Moore (1988), neither the central bank nor any other authority is able to determine the total stock of money in the economy. Of course, money is created somewhere and has a specific origin. But it does not come from outside the system nor does it fall from heaven. It is created during the course and out of economic activity itself. Money is created *ex nihilo* when banks provide loans to firms and customers. Every loan represents a purchasing power, and endogenous money is credit money. Each unit of money corresponds to a debt in the same amount (see, for example, Rochon, 1999, pp. 8–9).

Banks grant loans to borrowers if the latter fulfil the criteria of a creditworthy customer. Being aware of this fact and given that money is the result of the credit creation process, it is evident that the stock of money is driven by the same forces that drive the demand for credit.[3] Demand for credit depends crucially on the state of the economy. The stronger economic activity and the better the future perspectives, the more credit is demanded in order to finance activities. As an important first result, money is demand-led (Moore, 1988, p. 19).

The central bank may try to raise the stock of money by supplying reserves with easier conditions to commercial banks, which transmit this policy by enlarging the volume of their loans. However, the effective quantity of money only increases if there exists a demand for credits. If future prospects are bad and investment activity lies down, no credit is needed and monetary policy becomes nearly powerless. The 'money market' thus features a demand function but not a clearly identifiable supply function, since money supply is itself demand-driven (Moore, 1988, p. 19).

But the endogeneity of money gives rise to the fact that the central bank cannot determine the money stock by a policy decision. It can merely regulate short-run interest rates, which represent an exogenous variable (Lavoie, 1984, p. 777). The short-run rate affects the price of loans and therefore impacts on the profitability of investment projects. But it cannot influence the demand side of credit creation, because that is determined by economic activity. It is sometimes argued that the interest rate is somehow also endogenous, because monetary policy can in substantial part be a reaction to the state of the economy (Woodford, 2001, p. 232). However, short-run rates of interest do not automatically respond to economic developments but rather remain being set by the central bank. Hence, their exogeneity is still given (Gnos & Rochon, 2007, pp. 380–381). Taken together, the finding is that demand-determined money and exogenous short-run interest rates do not necessarily correlate in contrast to the notion of exogenous money.

The setting of the interest rate has been, and sometimes still is, the issue of a big debate among proponents of endogenous money. According to one faction, called structuralist and strongly influenced by Minsky (1957), the interest rate curve is an increasing function in the volume of loans given to borrowers (see, for instance, Dow, 2006; Palley, 1991, 1996; Pollin, 1991). Investment grows in the course of economic activity but the central bank does not ease interest rate conditions infinitely. Hence, commercial banks have to manage their liabilities and therefore raise lending rates of interest. Since the central bank limits the availability of reserves, there is not only a demand constraint but also a supply constraint on credit creation (see Le Heron & Mouakil, 2008, pp. 421–424).

A second faction, the so-called horizontalists, assesses the interest rate curve as a horizontal line (Kaldor, 1982, 1985; Lavoie, 2006a; Moore, 1988). The central bank sets the rate at a certain level. At this level, the whole demand for reserves is accommodated. The monetary authority can determine the price of credit but it cannot freely decide to limit supply when demand is high. If it refused to provide all required reserves, it would have to fear that banks get into financial distress. Turbulence may affect the real economy negatively. In analogy to the central bank, commercial banks grant loans according to demand and under the condition that the borrower is creditworthy (Lavoie, 2006a, p. 24; Rochon, 2006, pp. 171–173). The thought that the central bank behaves in a fully accommodative manner is an important feature and in strong contrast to the theory of exogenous money as will be seen later (Rossi, 2008, pp. 189–190).

There are other theories of endogenous money like the circuitist approach (see Gnos, 2003, 2007; Parguez & Seccareccia, 2000) or the theory of money emissions (see, for instance, Cencini, 2005; Rossi, 2003, 2006b, 2009b; Schmitt, 1960). They focus on the circulation of money from the beginning when a loan is granted, through production of output, to income destruction and repayment of the loan.[4] Rather than a contrast to the structuralist and horizontalist conception of money, they are, despite differing focus and some theoretical dissension, related as they share the theory of endogenous money (Rochon & Rossi, 2003, pp. xxv–xxvi). Further, there is the New Consensus approach to monetary policy. As mentioned, it treats money as endogenous but it does not provide a distinct theory of endogenous money. It should be noted that the importance of the horizontalists has increased in the past. Leaving the New Consensus approach aside, horizontalism can now be seen as the dominant view within the endogenous-money paradigm (Rochon, 2007, pp. 4–6).

If money is demand-driven, the concept of the money multiplier does not make any sense. Once the central bank has set the short-run interest rate at a certain level, borrowers demand a certain volume of credit that they need in order to finance their activity. If economic activity is weak for whatever reasons, few loans are issued, the stock of money remaining low. An ever increasing supply of reserves by the central banks helps very little in this situation. Thus, it is not money reserves that drive the stock of money by means of a multiplier. It is rather the demand-determined stock of money that drives the amount of reserves that banks need to hold (see Lavoie, 2003, pp. 523–524). Even if the multiplier idea is potentially confirmed by empirical investigation, it does not have any causal meaning, since it is *a priori* wrong from an analytical point of view (Moore, 1988, p. 85).

In a world of credit money, all money is mirrored in debt in the same amount. Loans are issued to finance investment. The newly created money enters into circulation through the purchase of equipment, the employment of labour force, direct consumption or the purchase of financial assets. Sooner or later, this money becomes someone's income. It is deposited in a bank account and functions as savings. By logic of accounting, the deposits equal the issued loans. This is an important aspect of endogenous-money theory: first, investment and savings are always equal as a rule of accounting (see, for example, Cencini, 2003a, pp. 304–312). Second, causality goes from investment to savings and not the other way around. From a macroeconomic point of view, investment determines savings, that is, loans create deposits (Howells, 1995, p. 90; Moore, 1988, pp. 3–4; 1989, p. 55). This is not a question of personal opinion but comes out of the logic of double-entry bookkeeping, which holds independently of the number of commercial banks or whether or not there exists a central bank (Lavoie, 2003, p. 506). Neoclassical arguments according to which too low an interest rate reduces deposits such that banks are constrained in issuing loans appear as rather fallacious (see, for instance, Creel et al., 2013, p. 10).

With money endogeneity, money is the result of economic conditions. It is not money that determines nominal output through its possible short-run and long-run effects on real output and the price level. It is economic activity itself that induces a stock of money needed for all transactions. Hence, as particularly highlighted by Davidson and Weintraub (1973, p. 1117) in a wage-bargaining model – even without explicitly assessing the endogeneity of money – money is not a cause but an effect. And as money creation intimately corresponds to the monetary requirements of the economy, there basically cannot exist an excess money supply. Moreover, since the stock of money increases in response to the rate of inflation rather than the other way around, it is for this very reason that money does not cause inflation (Arestis & Sawyer, 2003, p. 9). This does not mean that monetary policy cannot have any influence on inflation. A low policy rate of interest can lead to strong credit creation. If thereby induced high demand exceeds production capacities, the price level increases.[5] Especially in

Kaleckian models, capacity utilization is argued to be below unity (Lavoie, 2014, p. 360). While it is often agreed on spare capacities in the short run, considerations about long-run capacity utilization differ (Lavoie, 2014, pp. 387–390). Hence, it is to a large part an empirical question and depends on the state of the business cycle or the degree of competition in the market. As long as production capacities are not fully utilized, the source of inflation lies in cost-pushes coming from higher commodity prices, more expensive imports as a consequence of a depreciated exchange rate, wage bargaining, and so on (see, for example, Kaldor, 1985, p. 10; Rochon, 2004, pp. 8, 19). Further, expansive monetary policy can also be linked to a falling price level. The so-called 'Gibson's paradox' is dubbed as 'one of the most completely established empirical facts within the whole field of quantitative economics' by Keynes (1930b/2011, p. 198). The paradox may get a couple of explanations in our specific context. One the one hand, a price puzzle might be present in Gibson's paradox to which we will refer in the empirical analysis. From a theoretical point of view, however, there is a more interesting idea. Interest payments represent a share in production cost of corporations. The lower the interest rate, the lower the production cost, which transmits to lower prices (Sawyer, 2002a, p. 42). In the same way, expansive monetary policy may accelerate economic activity. Previously installed fixed capital is more charged to capacity. More units of output share the cost of fixed capital and thereby the price of each unit falls (Arestis & Sawyer, 2009, p. 42). On these grounds, monetary policy has an influence on output but does not necessarily raise the price level. The creation of money itself is not inflationary. Moreover, by influencing investment, monetary policy may also affect the supply side and hence the productive capacity of the economy, which again reduces the inflationary effects of monetary policy (Colander, 2001; Lavoie, 2006b, pp. 178–181).

The theory of endogenous money in its proper sense rejects the idea of a natural interest rate. The latter is indeed unobservable and thus hypothetical (Smithin, 2013, pp. 244–246, 252). To apply it to monetary policy, it can only be estimated by crude approximation (see, for instance, Taylor, 1993, p. 202). As such, the natural interest rate is merely an assumption. Since it is just part of a theory without any evidence of real existence, there is no reason to accept this existence. This has far-reaching consequences that will also be important for our further investigation.

To sum up, there are four central features of endogenous money: it is demand-determined; investment is always equal to savings; the emission of money is non-inflationary by its nature; and there is no natural interest rate. Moreover, there is the dominant view within endogenous-money theory that the central bank behaves in an accommodative manner after having set the short-run interest rate. Taking these features together means that the quantity of money can, in principle, rise and fall by large amounts without having any causal effects. When the demand for credit is high, loans are issued. The accommodative behaviour of the central bank provides demanded reserves. The money stock grows without necessarily giving rise to a higher inflation rate. Different stocks of money can correspond to different economic outcomes. These characteristics reveal that, in contrast to the exogenous-money view, money is something completely different from a commodity. It is not scarce. Given the rate of interest, money is available in abundance. The only condition to create it by issuing loans is the creditworthiness of the borrowers appraised by the banks.

It is not easy to assess whether money is more often acknowledged in the literature as exogenous or endogenous. As exogenous money and the quantity theory take an important part in neoclassical economics, it may be argued that it is as dominant as neoclassical economics itself. However, the New Consensus as a new Keynesian or new neoclassical synthesis approach is also very close to the dominating paradigm but treats money basically as endogenous (although it does not explain this on theoretical grounds). Moreover, exogenous money as such does not have the same importance for neoclassical theory that endogenous money has for, by way of example, post-Keynesian economics. Neoclassical models tend to be constructed in merely real terms. Money is normally introduced in a further step as a kind of commodity. In that way, money is predetermined as exogenous. Endogenous-money theories, however, include money as a crucial part of their analysis right from the beginning. Money is more strongly emphasized as an integral part of any economic activity. This arises from the fact that monetary theories tend to incorporate all economic processes from financing production to final consumption. They refer to 'economies of production' and criticize neoclassical 'economies of exchange' where only the phase of exchange in the market is enlightened and the origin of money remains unexplained (Graziani, 2003, pp. 25–26). From this point of view, endogenous money might be put to the forefront.

Definitely responding to the question by empirical analysis is not an easy task but tends to come out in favour of endogenous money. A correlation between the stock of money and nominal income can be, and has effectively been, seen as evidence for the causal link from money to prices. Critics do not necessarily deny the existence of such correlations (see Kaldor, 1970, p. 5). But as anywhere, correlation is not the same as causation. Simple correlation therefore does not yield a clear result because whether money is exogenous or endogenous depends on the direction of causality. A study about the evolution of money over time by Schularick and Taylor (2009) finds that while monetary aggregates developed in proportion with total bank assets before World War II, they did not thereafter. The authors do not postulate money endogeneity by a theoretical analysis. However, they find that credit creation and thus money creation in a proper sense are not driven by an exogenous-money supply. Howells (2006) provides an overview of empirical literature about endogenous money and finds clear support for its existence. Furthermore, he argues that money is accepted as endogenous in the practice of monetary policy implementation even if it is not always emphasized theoretically as such (ibid., p. 65). Yet, as an example from practice, economists from the Bank of England explain the mechanism of money creation through the issue of loans by commercial banks and thus confirm the endogeneity of money from an analytical point of view (McLeay et al., 2014, pp. 14–25). So all in all, it seems fair to say that if the theory of endogenous money is not yet supported by a majority of theoreticians, its importance is at least growing fast.

The confrontation of the two paradigms yields some first insights that will be useful for our analysis. Within the framework of exogenous money, monetary policy leads to price changes without having a sustaining effect on quantitative variables like output, production capacities or employment. In the long run, the only suggestion for the monetary authority is to aim at a path following the evolution of the natural interest rate or the corresponding real business cycle. In contrast, endogenous money implies that monetary policy does not inevitably impact on prices but may influence quantities. The latter takes place under the condition that there exist sufficient demand and spare capacities (or capacity-enlarging investment) such that there is an effective reaction of borrowing in response to a change in monetary policy. Active monetary policy deviating from a hypothetical equilibrium path does not categorically have lasting destructive effects. Thus, under the endogenous-money paradigm, the amount of money is allowed to react much more elastically to changes in the interest rate level. Under a given monetary policy stance, the stock of endogenous money can basically grow interminable as demand for credit increases. This finding strongly contrasts the view of an exogenously determined – and hence limited – stock of money. The elasticity of endogenous money and its nature as a reaction to, rather than a cause of, economic activity will be a crucial feature in the examination of the role of financial markets in the next section.

2.1.2 The Role of Financial Markets in Monetary Policy

Financial markets and the role they play in light of monetary policy is an issue that has rarely been debated as intensely as in the years since 2008. This section introduces two different views of financial markets. The first is the neoclassical view that dominates nowadays. However, as will be shown, there are many reasons for criticism. This is why a second, alternative view of financial markets is introduced. It is based on endogenous money and is used thereafter in the remainder of our investigation. Emphasizing the role of financial markets is important for the subsequent analysis of the crude oil market.

The neoclassical view
According to the neoclassical view, economic outcomes are determined by consumers' utility function and firms' production function.[6] They are of a purely technical nature and do not take into account any endogenous monetary dynamics. In the case of perfect competition, not even social relationships, like the balance of strength between employers and workers, play any role. They are only considered when market imperfections or nominal rigidities are introduced, as is the case in new Keynesian models, for instance. But still then, these features are treated in a rather static way. Thus, in the long run, the economy is driven exclusively by real forces, the so-called fundamentals. According to the theory introduced above, these fundamental forces determine the natural interest rate. At this rate, the only sustainable one, the economy is in equilibrium. Hence, money is added as exogenous to the basic real forces and treated like a commodity (see Bénicourt & Guerrien, 2008, p. 241).

Even though the theoretical background is missing a clear idea of money, it does not imply that money is completely meaningless. At this point, financial markets should be examined. Neoclassical theory postulates two main hypotheses in this respect. The first one is the rational expectations hypothesis (see, for instance, Lucas, 1972; Sargent, 2008). As its notation says, agents build their expectations about the future on rational grounds. They learn from results in the past and adapt their expectations in order to avoid errors. This is an ongoing feedback process. As a consequence, agents forecast their variables of interest without bias. All forecast errors are only random. The second postulate is the efficient market hypothesis, which can be considered as an application of the rational expectations hypothesis. According to the idea of efficient markets, all available information about an asset is incorporated in its price (see, for instance, Fama, 1970, p. 383; Jones & Netter, 2008). For example, when a new promising production technology enters the equipment of a company, higher future dividends can be expected. The stock price of this firm increases. The hypothesis is equivalent to a no-arbitrage condition. Any new available information is immediately reflected by the corresponding asset price. Speculative profits are not completely excluded but they are random. Every reaction of a price to new information brings random profit to some agents. However, random profit is the requirement for the no-arbitrage condition, as it eliminates the possibility of any further, systematic profit. This corresponds to the well-known saying that it is impossible to beat the market. Many studies reveal that professional investment managers are not able to reach systematically higher returns than the average of all stocks in the market (Malkiel, 2003, pp. 77–78).

It is in the credit and capital market where capital demand and supply meet. Financial markets are therefore useful for improving the efficiency of capital allocation. Especially when capital is made marketable in the form of stocks, bonds and other securities that are traded in primary and secondary markets, it is most likely that capital finds its optimal investment opportunity. The more liquid financial markets are, the better and faster price discovery is achieved (see Jones, 1999, pp. 1506–1507). In this sense, a growing volume of transactions in financial markets can be interpreted as an evolution towards perfectly working market mechanisms.

The economic fundamentals as the independent and only driving forces of economic outcomes on the one hand, and the financial sphere with exogenous money on the other, are given. The efficient market hypothesis is a crucial tool for understanding the relationship between them. It makes all movements of asset prices directly depend on changes in fundamentals. For instance, considering the stock market, we know that high stock prices tell us about higher profit expectations, themselves a sign of well-working fundamentals. Thus, financial markets are a reflection of the real economy. Causality goes from fundamentals to financial markets. Financial assets do not have any intrinsic productive force, so they are not a key determinant of fundamentals. That is why there is no causality from financial markets to the real economy. Deviations from this standpoint are due to market imperfections. These may be rigidities or costly information as will be seen below.

Coming from the quantity theory of money, a change in money supply leads to a proportional change in the price level over the long run. Following the further evolution of neoclassical theory after the 1970s by including new Keynesian market imperfections and real business-cycle approaches, monetary policy becomes partially more effective but the basic proposition of the quantity theory with exogenous money still holds. The hypothesis that expectations are rational is not affected by any action of the monetary authority. Thus, the efficient market hypothesis remains valid, even if one allows for some imperfections. The relationship between fundamentals and financial markets does not change in the course of active monetary policy. In the long run, the latter has only a nominal effect on the economy. Consumers' utility function remains the same owing to their rational behaviour. Production technology does not change either. Hence, real forces are not influenced. It is still real variables that determine real profits and it is the expectation about real profits that financial markets react to. As a corollary, since monetary policy does not affect the fundamentals, it does not have an impact on financial markets performance. Financial volatility is the symptom of real business cycle fluctuations. In cases of crises or even depressions, the sources do neither lie in financial speculation nor in distortions inherent to the economic system. They are triggered by exogenous shocks in productivity, labour or capital, which are in general caused by misled public interventions (Kehoe & Prescott, 2007, p. 15). Woodford (2002, p. 87) even argues that efficient financial markets support the conduct of monetary policy without distorting efficient capital allocation, to wit, without affecting fundamentals.

Nevertheless, empirical evidence is found in favour of a significant effect of monetary policy on asset prices, in particular stock prices (see, for instance, Ehrmann & Fratzscher, 2004; Rigobon & Sack, 2002). There are several explanations to reconcile these observations with neoclassical theory. It can be due to imperfections with respect to the efficient market hypothesis or to short-run effects on economic fundamentals, which are then reflected by a corresponding response of stock prices. Furthermore, time lags are certainly larger in the real producing economy than in financial markets. A change in real stock prices in the course of monetary policy might then be explained by the fact that the changed money supply impacts first on financial asset prices and is offset by a corresponding change in consumer prices later. This corresponds to the results found by Bordo et al. (2007, pp. 18–23). They argue that for the period since World War II, stock market booms have occurred when interest rates and inflation rates were low. Booms tend to bust when the rate of inflation increases. From this exogenous-money view, monetary policy necessarily cannot have a lasting effect on real stock prices, because it is clear from the beginning of any policy activity that it will affect the general price level in the long run. Eventually, the general price level increases for the same reasons as stock prices: the increased supply of money has exerted its effects; the initial impact of monetary policy is neutralized.

To sum up, money is neutral in its relationship to the real economy as well as to financial markets. Some conditions have to be satisfied for this approach to be useful. First, money neutrality should not be distorted too much by market imperfections and nominal rigidities, as we already noted in the previous section. More distortions mean less money neutrality. The second, even more crucial condition is the stability of the relationship between fundamentals and the financial sphere. If expectations are rational and markets are efficient according to the two hypotheses at the core of neoclassical economics, this relationship should be stable. It is true, then, that financial markets do not feature any independent dynamics that could spill over to fundamentals.

However, many economists within the neoclassical paradigm disagree about the efficient market hypothesis on empirical grounds (see Summers, 1986). Besides doubts as to whether a fully rational behaviour is realistic from an analytical point of view, there is the problem of measurability. The so-called joint-hypothesis problem is recognized among both proponents and critics of the efficient market hypothesis. For instance, Fama (1991, p. 1575), a founding economist of the efficient market hypothesis, states that 'market efficiency per se is not testable'. To define rational behaviour and efficiency, a theoretical model is needed. Deviations from such a model found in empirical tests, then, are due either to existing market inefficiency or to the incompleteness of the model. But it is not clear where to attribute the measurement errors.

Grossman and Stiglitz (1980) articulated an early important criticism. In a world with imperfect, costly information, informed agents face arbitrage opportunities. The information spillover to non-informed agents through price changes is imperfect. Arbitrage profits are therefore a necessary condition for an efficient market. They give agents an incentive to produce costly information, which is again necessary for the market to find the equilibrium price. As is widely recognized, this distortion of the efficient market mechanism is more realistic than the assumption of perfect information.

Shiller (2003, p. 86) argues that if the efficient market hypothesis held, stock prices should be equivalent to expected dividends, that is, the present value of dividends. However, data show that while expected dividends follow a stable trend throughout the twentieth century, stock prices are much more volatile and deviate over long periods from expected dividends. These observations of seemingly irrational behaviour gave rise to intensive research in behavioural finance (see Olsen, 1998). Investors may not only react to information about fundamentals but also be just feedback traders who react to past price developments (Shiller, 2003, pp. 91–101). Proponents of the efficient market hypothesis tend to share the view that potential misguided price changes caused by feedback traders are offset by well-informed agents who benefit from this arbitrage opportunity. But even this view is challenged. There are models where rational behaviour has a destabilizing rather than a stabilizing effect on stock prices. For instance, in De Long et al. (1990), rational investors buy assets and thereby raise prices. They know that feedback traders will follow them and raise prices further. Rational investors can now sell at a higher price. Hence, they exploit the arbitrage opportunity that they have themselves created before.

In general, these critiques originate from economists who do not leave the neoclassical paradigm, but instead broaden it. The more imperfections there are, the longer it takes for asset prices to adjust to fundamentals and the more influence monetary policy can therefore have. It is questionable as to which degree of market imperfections or irrational behaviour it is adequate to assume the correctness of the efficient market hypothesis. The claim that asset prices will eventually revert to their fundamental value loses its significance the more imperfections distort the equilibrium pattern. The short run may effectively become very long. During the period when asset prices deviate from real economic conditions, the causality between fundamentals and financial markets can become bidirectional. Asset prices then are not a simple reflection of real forces but are also driven by other factors. Hence, financial market performance appears to fundamentals as a partially exogenous variable: it is influenced by fundamentals but may also have an impact on fundamentals.

An alternative view
Neoclassical theory provides an explanation of financial markets and the role of monetary policy that is basically consistent. To arrive there, however, strong assumptions are required. They are useful for the completeness of theory but not necessarily for the description of economic reality. As an example, the existence of stock market bubbles is partially acknowledged by proponents of the efficient market hypothesis, for instance during the internet boom at the end of the 1990s. But they still hold up their support of the hypothesis since there was no evidence of any systematic arbitrage opportunities during that bubble (see, for instance, Malkiel, 2003, p. 75). So, if bubbles occur even if markets are efficient, it remains to ask what the efficient market hypothesis should be useful for. The fundamental question from the beginning was and still is whether financial markets perfectly reflect the real forces of the economy or whether they feature their own dynamics. Asserting the correctness of the efficient markets hypothesis is considered as sufficient evidence that financial markets follow a neutral pattern. Yet, evidence for both efficient markets and bubble-building jeopardizes the comfortable findings of neoclassical financial theory. This is in fact argued to be a central weakness of the efficient markets hypothesis: market efficiency on the one hand and the condition of no arbitrage in financial markets or the non-predictability of asset prices on the other are not at all the same thing. However, they have been taken to be the same thing, or at least as two inevitably connected issues, for decades by proponents of the efficient markets hypothesis (see, for example, Guerrien & Gun, 2011, pp. 25–26).

In what follows, we emphasize an alternative view of the role of financial markets in monetary policy. It is a view that does not rely crucially on the microeconomic assumptions of how individuals behave. Moreover, we do not assign all potential deviations of nominal variables from fundamentals to rigidities. Rather than a wedge in the otherwise smoothly working economy, price and wage rigidities are a basic characteristic of capitalist production. With complete flexibility of all variables, it would be impossible to plan production (see, for instance, Cottrell, 1994, p. 591). The perspective we adopt in this regard is a macroeconomic one. Building a macroeconomic theory by simple aggregation of individual actions usually requires strong assumptions and simplifications in order to allow any aggregation at all. Human behaviour and microeconomic market mechanisms are extremely complex. Summing up individuals and their economic actions in a mathematical way ignores interactions between them. Macroeconomic results can thus get rather far from reality. This phenomenon, which is well known in economics, seems also very likely to occur in financial markets: the famous fallacy of composition (see, for example, Lavoie, 2014, p. 17). The above presented hypotheses of rational expectations and efficient markets are symptomatic. As Cencini (2003b, p. 8) puts it, perfect competition and rational behaviour are the necessary assumptions allowing the distinction between micro- and macroeconomics to be only 'a matter of size and not of substance'. However, rational behaviour from an individual point of view is not necessarily rational nor is it necessarily efficient from a macroeconomic perspective. In a financial market boom, investors may behave fully rationally if they speculate on the irrational behaviour of others.[7] They believe that prices will rise because of other investors' enthusiasm and therefore purchase assets. If a sufficient number of investors act like this, prices will effectively increase. Perhaps it is the case that no one at all behaves irrationally, because what has been expected before is reality now. Moreover, it is difficult to obtain a higher than average market return, that is, it is unlikely to beat the market. Nevertheless, asset prices have probably moved away from their real intrinsic value. A bubble might build up and burst at any time.

Investors build their expectations about the future. But the future is uncertain. This view is related to the principle of uncertainty as developed by Keynes (1936/1997, pp. 161–162). It says that uncertainty is not just a set of possible outcomes with an exact realization probability for each. The future is fundamentally different from the past and the environment is too complex and in permanent change. Mathematical probabilities as a tool to make uncertainty more certain are therefore doomed to failure (Keynes, 1937, pp. 213–214). An investor has to build a belief of what other investors believe, while the belief of other investors is built on what they believe that the other investors believe that they believe, and so on. This is an important mechanism for how investors build their short-run expectations in an uncertain economy (Keynes, 1936/1997, pp. 154–158). This means that macroeconomic outcomes in financial markets cannot be traced back linearly to the behaviour of individuals because of dynamic interdependences. It is not useful to base the theory on the assumption that individuals behave either rationally or irrationally. Whether the former or latter holds true, uncertainty does not allow deriving a macroeconomic theory from a simple summation of suggested individual behaviour. In Cencini's (2003b, p. 13) words, the rational expectations hypothesis has to be rejected owing to 'the logical impossibility to derive macroeconomics from microeconomics'. The conclusion for the efficient markets hypothesis is analogous.

In an economy with uncertainty, expectations and behaviour of individuals can shift macroeconomic variables in principle anywhere. This gives rise to saying that in a world of uncertainty, the conception of equilibrium to which the economy converges is meaningless (Weintraub, 1975, p. 535). Such equilibrium does perhaps not even exist and if it does exist, it is hardly possible to determine its level. Assuming a doubtful equilibrium path of the economy may impede a realistic analysis.

The macroeconomic approach in this chapter is based on the principle of endogenous money. The demand-determined creation of money provides large flexibility for the financial system. In combination with an economy of uncertainty, endogenous money brings about a more autonomous financial market evolution. Given the expectations of investors, endogenously created money influences asset prices. In contrast to the neoclassical paradigm, financial asset price changes therefore do not necessarily have to correspond with changes in real forces. These impacts may even last in the long run, because there is no equilibrium to which asset prices should converge together with the real economy. Since financial markets are not just a simple reflection of fundamentals, this approach allows for a bidirectional relationship between the real economy and the financial system. Financial markets can have a positive impact on the producing economy by providing liquidity. But they can also build bubbles that may be disastrous at the time of their bust. Davidson summarizes these characteristics in the allegory of the 'double-edged sword of financial markets' (Davidson, 2002, pp. 104–105). If flexibility is an outstanding feature of financial markets, fragility must be added as another property. It is in this environment of uncertainty that Kaldor's (1939, pp. 1–2) definition of speculation applies: financial investors make use of expected price changes and thereby may themselves affect the price.

Even though money is not considered as neutral and financial markets are not assumed to simply follow the pattern of fundamentals, it should not be claimed that the real economy and financial markets are entirely unrelated. The performance of production is the basis of present and future profits. Agents' expectations therefore depend on fundamental data. However, this relationship is neither linear nor does it follow any predetermined time pattern. Fundamentals and the financial system affect each other and can, beyond this, be altered by other forces. Specifically, monetary policy can influence the connection. It acts as an exogenous force by setting short-run interest rates. This exogenous force has the – imperfect and sometimes quite limited – capability of distorting the relationship between fundamental and financial variables. The means of distortion is endogenous money.

When a commercial bank grants a loan to a company, an asset is created. The new money goes to a bank deposit. From there, it will be used to buy production equipment. In a world of endogenous money, loans create deposits. While debt represents a financial asset for the lender, purchased real goods are real assets. Money has a payment function. Financial assets are the counterparts of bank deposits. And money is, owing to its endogenous nature, a measure of produced output and thus it embodies purchasing power by raising a claim on output (Rossi, 2008, pp. 38–39). The principles of double-entry bookkeeping should be emphasized for the present purpose.

Table 2.1 shows the balance sheet of an individual company. Its assets are on the left-hand side. The right-hand side represents financial assets possessed by those investors who fund the company equipment by providing capital. From the view of the company, however, the right-hand side is not assets but liabilities. In this respect, equity can be considered as a debt

Table 2.1   Balance sheets of a production company and a commercial bank

Company

Assets	Liabilities
Cash	Borrowed capital
Bank deposits	Corporate bonds
Equipment, real estate, etc.	Equity

Bank

Assets	Liabilities
Loans	Deposits
	Advances from the central bank
	Equity

Source: Author's elaboration.

of the company to its owners. For a bank, the balance sheet looks basically the same. However, its assets are the company's borrowed capital as well as optionally corporate bonds and the company's equity, in the form of stocks. Its liabilities are bank deposits held by the public, equity, and, in the exemplary case of Table 2.1, advances from the central bank.[8]

At first, let us consider the case of a pure production economy. The financial market exists merely as a result of the fact that banks create money through the issuance of loans. They are needed to finance production. The quantity of money depends on production. Money is therefore a measure of output. The existence of the financial market is a direct cause of real activity rather than a proper autonomous force. In Table 2.1, equipment, real estate, property rights and the like are real assets. The liability side consists of financial assets. The difference between real and financial assets is made up by money in cash and deposit form. In our production economy, this money can be spent to purchase a corresponding share of output. Alternatively, given that output is already produced and sold such that cash and deposits represent sales earnings, money can be used to repay debt. When output is destroyed by purchase and consumption the money stock decreases by the same amount through debt repayment. Money corresponds to output in a stable relationship, since it is caused by output. Therefore, debt and real assets are equal from a macroeconomic point of view. This is shown in Table 2.2, where debt entirely consists of bank credit.[9] The equality is measured in units of money.

In the case of a pure production economy, bank loans are the simplest expression of the relationship between the borrower and the lender. Credit is granted and paid back at the agreed time. Nothing happens in between. Yet, in reality, debt is securitized to a large degree. Stocks and bonds are not only a borrower–lender relationship like bank loans used to be, they are also tradable papers (Lavoie 2003, p. 518). An investor who buys stocks becomes the lender of somebody he probably does not even know. As soon as debt becomes tradable, it is measured twice. First, it has its original, so-called nominal value, which is the sum that has been borrowed from the lender. Second, it has a market price, depending on demand and supply. This gives rise to the existence of a market for financial assets. It is through

Table 2.2   Balance sheet of the production economy

Assets	Liabilities
Real assets	Bank credit

Source: Author's elaboration.

securitization of debt that financial markets get an autonomous role with their own dynamics, which have the potential to influence other markets.

To begin with, the supply side and demand side of the financial asset market are highlighted. Financial assets are supplied by producing companies. When a company plans new investment that is not financed out of its own means, it demands a loan at its commercial bank. Alternatively, it issues new stocks or bonds. While the firm requires capital on the one hand, it provides a debt title either in the form of a loan or security papers on the other. The capital is used to purchase real goods like equipment, real estate and more. In this way, the company builds the potential for future profits. Parts of these profits are distributed to the owners of financial assets in the form of interest and dividends.

Financial asset demand is generated by investors. The term 'investors' includes commercial banks as well as (private) non-bank investors. In contrast to private investors, banks have access to central bank reserves. Hence, they play the role of money creators. Financial asset demand depends on two main factors: expected profits of assets and liquidity preference of investors. Profits are composed of future security earnings, that is, dividends and interest as well as the change in the market price of financial assets. In neoclassical theory, the latter component is a simple reflection of the former. The higher the dividends, the higher the fundamental value of the security and the higher therefore its price. In the alternative view, financial asset prices have their own dynamics owing to uncertainty. Investors build beliefs about the others' beliefs, which influence individual asset demand and eventually total demand. Demand affects prices. Price changes again impact on future demand.

Liquidity preference means that investors weigh the advantages of having wealth in a liquid form against the return prospects of investing it in a riskier, less liquid asset.[10] The idea of liquidity preference is closely related to the existence of uncertainty in the economy (Lavoie, 2014, pp. 238–250). The higher the liquidity preference, the less investors are willing to provide capital for investment, that is, to purchase assets. When risk in financial markets is suggested to be high, investors substitute financial assets with money, the most liquid form of wealth. In times of low liquidity preference, demand for securities increases. Changes in liquidity preference are related to changes in expected profits. But they are not synonymous, as we will point out when discussing them in relation to monetary policy.

In a given moment, a firm faces a certain funding requirement which determines its amount of debt. A fraction of the debt is securitized. Hence, we are not in a pure production economy but in an economy with financial markets where financial assets are traded. While financial asset supply is given, financial asset demand is assumed to increase. Securities prices rise.

Table 2.3   Balance sheet of an economy with securitization

Assets	Liabilities
Real assets	Non-tradable stocks and bonds (nominal value)
Non-tradable stocks and bonds (market price = nominal value)	Tradable stocks and bonds (nominal value)
Tradable stocks and bonds (market price)	Bank credit

Source: Author's elaboration.

It is further assumed that demand augmentation is strong enough such that prices are above their nominal value. Demand cannot become effective out of nothing. It is reflected in a corresponding amount of money, which is used to purchase financial assets. To create this money, new loans have to be issued. There is now a higher money stock but also a higher stock of total debt. However, production is still the same. Table 2.3 exhibits the total balance sheet of the economy. In contrast to the pure production economy in Table 2.2, total debt is now larger than total real assets. As a second difference, liabilities are not only credit owed to banks. There are, in addition, stocks and bonds, which are owned by investors. They appear therefore both on the assets and liabilities side of the economy. While companies have debt in the amount of the nominal value of stocks and bonds, investors possess them as financial assets at market price. The difference between market price and nominal value of the assets is caused by increased security demand, which is itself financed by new bank credit.

Higher demand in the financial asset market does not raise the rate of inflation in the economy, that is, the inflation of consumer prices, as long as the additionally created money remains in the financial sphere and is not spent on the real goods market. In the case of a falling volume of loans, financial asset demand falls and therefore triggers asset sales and further price drops. This is a financial crisis. The balance sheet in Table 2.3 grows shorter. For now, it is sufficient to say that in this kind of economy, as long as consumer prices are stable, total money cannot raise a claim on total output. A part of the claim is devoted to financial assets.

The supply and demand aspects of financial assets reveal that the notion of investment should be clarified. Corporations want to invest but it is not they who finance investment. They raise capital by borrowing or by issuing stocks and bonds. Financial investors are only interested in pursuing their investment strategies. Whether the investment leads to the installation of additional production capacities is of secondary priority from their point of view. In contrast, companies need capital in order to buy equipment, increase output and obtain profit. It is only in this way that investment has a direct effect on the real economy. Hence, we shall refer to 'real investment' when companies use it to buy production capital. On the other hand, we refer to 'financial investment' when investors purchase financial assets that bring them a certain annual return. Real investment corresponds to the supply side of financial assets while financial investment reflects the demand side.

Monetary policy impacts on financial markets in the alternative view

Given the basic anatomy of the financial asset market, we shed light on the role of monetary policy. Let us consider the case of expansive monetary policy. A fall in the level of the interest rate triggered by the monetary authority reduces the investment cost of corporations. They raise their demand for capital to increase investment. Additional bank loans are therefore issued. From the perspective of an individual firm, the emission of tradable securities depends likewise on the level of the interest rate even if mainly through indirect channels. Corporate bonds can be issued at lower cost because financial investors require lower remuneration in the face of a lower general interest rate level all over the economy. New stocks can be emitted at easier conditions for the same reason, too. Hence, expansive monetary policy raises the supply of financial assets (see Mishkin, 1996, p. 2).

The demand side is influenced by monetary policy in a similar way. A lower interest rate does not necessarily raise expected profits in the beginning. But it lowers liquidity preference. Banks can refinance themselves under easier conditions on the interbank market. For private investors, it becomes more lucrative to use their savings deposits to purchase financial assets. Bank loans and securities are less liquid and riskier than cash but they generate a higher return. For professional investors, more capital can be borrowed at a lower interest rate and be invested in securities. Hence, lower liquidity preference induced by an expansionary monetary policy raises demand for financial assets. Resulting price changes may boost profit expectations of speculative investors. As a repercussion, asset demand grows further. To briefly link this argument to the criticism of the efficient market hypothesis above, this outcome is possible whether investors behave rationally or irrationally. In a world of uncertainty, speculating on certain behaviour of other investors in the market can be absolutely rational. Higher demand for securities reflects higher demand for credit and thus larger money creation. It can again be reasonably assumed that banks are accommodating on the condition that borrowers are judged as creditworthy (Lavoie, 2006a, p. 24). Thus, demand for assets is, in principle, allowed to grow without limit as a reaction to expansive monetary policy.

It is both a theoretical and an empirical question whether nominal or real interest rates are relevant when arguing about the effects of monetary policy. Neoclassical theory clearly suggests the use of real rates and, consequently, considers nominal rates, at least partially, as irrelevant. Since monetary policy can only have nominal effects in the long run, it is only potential changes in the real rate of interest that can have lasting impacts. Indeed, Greenwald and Stiglitz (1987, p. 121) state that it is 'real, not nominal, interest rates that should matter for investment'. However, from the endogenous-money viewpoint, all money that is created defines a claim on output produced while output is itself measured by monetary units. Therefore, 'economic magnitudes such as prices, income, profits, capital, interest rates and so on are simultaneously monetary and real and cannot be determined separately either in purely monetary or real terms' (Cencini, 2003a, pp. 303–304). This is the argument we will apply in the remainder of the book. It is only nominal interest rates that can be measured. Real rates have to be calculated first and do not exist as an objective observable indicator. It is thus not wrong to rely on nominal rates.

The motives to purchase assets can be manifold. Most investors seek profits. Expansive monetary policy lowers their liquidity preference. Expectations of future profits and dividends increase. In such an environment, investors with profit purpose raise their demand for financial assets. Other investors want to diversify their investment portfolio in order to reduce risk (see, for example, Markowitz, 1952). In an environment with low interest rates, diversification can be achieved at lower cost; it is easier to finance a well-diversified portfolio. Efforts for diversification can lead to a spillover of price changes from one asset class to another. A third motive is the store of wealth. Possessing money in its liquid form brings some advantages. But money is also vulnerable to changes in the price level. When the monetary authority lowers interest rates, liquidity preference decreases while inflation expectations increase. Thus, transferring wealth to a less liquid but safer form, that is, financial assets, becomes more attractive (see, for instance, Berck & Cecchetti, 1985). In all these cases, expansive monetary policy raises the demand for assets.

But owing to the endogeneity of money, there does not exist a linear relationship between the setting of a short-run interest rate and money demand. This was already outlined in the first section. Hence, the degree to which monetary policy can effectively change demand for money depends on many factors related to the business cycle, market structures and investors' expectations. This same issue applies to the demand for financial assets: lower interest rates tend to stimulate asset demand but the magnitude of this effect depends on a multitude of factors. We will analyze this in more detail later on.

As a second specification, we should draw a distinction between different asset classes. Not all financial assets are the same. Their liquidity depends on the state of the business cycle. As argued by Davidson (2002, p. 105), in times of recession or stock market crashes, severe liquidity crises occur. Government bonds that are classified as safe tend to play the role of a secure haven in times of financial crises. Their liquidity is therefore more stable. In a stock market downturn, prices fall. Debt repayment and investors' bankruptcy make money demand fall, too. However, many investors prefer to hold government bonds rather than pure liquid money. Bond prices then may rise. Thus, there can be opposing trends in the evolution of stock market and bond market as, for instance, remarked by Keynes (1930b/2011, pp. 83–84). Higher bond prices are the counterpart of larger money demand in the economy and therefore a larger volume of credit as illustrated in Table 2.3. This backlash cushions the drop in the money stock when a financial crisis occurs. Conversely, money grows more slowly when the crash is over and financial markets start moving towards a new boom, because there is a tendency to shift away from safe havens, which lowers bond prices in the first stage. While this effect sustains the stability of the quantity of money, the volatility of prices may be even stronger. Investors exploit arbitrage opportunities by weighing returns and risk and moving to the assets where prospects are best. The effect of monetary policy on the market for financial assets may therefore be actually enhanced. When a lower interest rate exerts its stimulating effects on demand for securities at a given supply, price changes in the stock market may trigger substitution of stocks for bonds and raise prices further. Yet, this does not give rise to generally falling bond prices in stock market booms. Higher liquidity in the course of expansive monetary policy is also likely to raise bond prices in the medium run. Correlation between stock and bond prices may be positive or negative (as well as insignificant in between) depending on the state of the business cycle (Li, 2002; Terzi & Verga, 2006, pp. 1–2, 5–6). All in all, there are complex interactions between different classes of financial assets.

Another specification to be made is the identification of asset demanders and asset suppliers. While above asset supply was assigned to producing corporations and demand to banks and private investors, the matter is in fact not that clear. Companies supply financial assets by funding their investment. But they may as well become asset demanders. For instance, when profits are not fully distributed, they are often reinvested. This can take the form of investment in the company's equipment. However, in many cases reinvestment is realized by purchasing financial assets, either by the repurchase of the company's own stocks or by acquisition of other

securities in financial markets. The increasing weight of the financial
sphere in the economy has made the structure of financial institutions
rather confusing. Securitization has raised the number of so-called shadow
banks, that is, various forms of investment funds. They issue shares to
raise capital and use it to invest in financial markets. Hence, they are both
financial asset suppliers and demanders.

Recall that, in the neoclassical view, investment requires saving. Saving
increases and investment decreases in the interest rate. The intersection of
the investment and saving curves determines the natural equilibrium rate
of interest. This paradigm does not make sense in the presence of endogenous money: loans create deposits and therefore investment is always
equal to savings. Demand for capital determines the supply of financial
assets. Supply of capital, on the other side, represents demand for financial assets. It is credit creation that satisfies capital demand by generating
capital supply. Based on this logic, supply of financial assets is always
equal to demand for financial assets. This is a necessary and always valid
principle rather than an equilibrium condition. Issuance of loans, and thus
the volume of deposits, tends to be larger when the interest rate is low. In
the same sense, both financial asset supply and demand grow when the
rate of interest falls. The level of the interest rate influences the quantities
in the market of financial assets but it cannot distort the equality between
demand and supply.

Yet, the necessary equality of financial asset supply and demand does
not imply that the price of financial assets is stable. There are different
forces on the supply and demand sides of the market that exert their influence through changes in price. It may then be argued from a neoclassical
perspective that this is exactly an equilibrium analysis, which states that
the price is the tool to bring supply and demand together. To reply, first, it
always applies that what is sold by an agent is purchased by another agent,
and *vice versa*. The equilibrium condition is therefore always fulfilled.
Second, the equilibrium can, in principle, be placed at every price and
quantity level (see, for instance, Asensio et al., 2010, p. 9). Equilibrium is
thus everywhere and disequilibrium is nowhere. The equilibrium approach
loses its utility. This implies the rejection of Say's Law, which states that
all supply is met by an equivalent demand under profit maximization of
all individuals (Davidson, 2002, pp. 19–21). The thereby resulting general
equilibrium in real terms is predetermined. Yet, in reality, the economy is
a monetary one; there is uncertainty and expectation building. The cross
point realized within the output-price space is not a static one determined
by production and utility functions but crucially depends on effective
demand (see Davidson, 2002, pp. 21–25; Hartwig, 2006). In Keynes's
(1936/1997, p. 55) own words, 'the effective demand is simply the aggregate

income (or proceeds) which the entrepreneurs expect to receive [. . .] from the amount of current employment which they decide to give'. This means, in a few words, that production is determined by expected sales. Once those sales take place, the intersection of the supply and demand curves determines price and output finally realized. In a world of uncertainty, the price thus cannot be predetermined and it can, in principle, take any value. With respect to financial assets, uncertainty is enhanced further. In contrast to a pure production economy where credit creation corresponds to the funds needed for production, there is no natural limit when speculation is introduced. Investors may raise their investment funds and raise demand for stocks without contributing to production. Therefore, supply and demand sides of the market can be subject to quite different and independent changes. The conclusion that supply and demand are therefore unequal in the end is nevertheless wrong. The only ever existing thing in reality is the cross point where supply and demand are necessarily equal in, notably, monetary terms.

The equivalence of financial asset supply and demand is expressed in the following simple formula:

$$I = q * p \qquad (2.1)$$

where I is the total amount that financial investors invest in financial assets, that is, it represents the demand side; q is the number of financial assets emitted by corporations; p is the price of financial assets. Financial investment I is always equal to real investment q at its current price p. This equivalence allows distinguishing between quantity and price effects. Taking again the same example, an increase in the demand for financial assets I can either translate into an equal growth in real investment q or in an increase of price p. A higher q is tantamount to the emission of new securities by firms. They imply an investment in the same amount in new equipment. If firms do not react to higher demand for financial assets, asset prices necessarily have to climb because assets are scarce.

The central questions to assess the role of monetary policy are the reactions of each financial asset supply and demand to a change in the interest rate. As shown, both respond positively (negatively) to expansive (restrictive) monetary policy. It is not *a priori* clear, however, whether financial or real investment changes more strongly when the rate of interest changes. Financial market performance is determined by the relative strength of the responses of the supply and demand sides. If expansive monetary policy leads to powerfully widening money creation because investors raise their demand for financial assets but generate only a weak emission of new financial assets by firms, securities prices increase. Conversely,

if firms start issuing a large number of new financial assets while financial investors raise their investment funds by a smaller amount, prices
fall. Supply is likely to rise when demand is high or even when it is just
expected to be high. Similarly, a high supply of new financial assets at a
low price may attract financial investors and encourage them to raise their
demand. These are the conventional feedback mechanisms of demand
and supply.

Financial asset supply is enhanced by expansive monetary policy
because investment costs decrease. The magnitude of the enhancement
depends on the profitability of additional investment. Expenditures for
equipment have to be reimbursed by higher sales quantity. Moreover, the
latter must be sufficient to cover interest cost plus a profit margin that
goes either to the holders of stocks or to the company in the case of bank
credit and corporate bonds. Ample sales are only guaranteed if sufficient
demand is assured. Effective demand includes only demand for real goods
and services but not for financial assets. Its sufficiency is a condition to be
fulfilled if monetary policy should be effective in stimulating the economy
through investment. On the other side, investment expenditures contribute to higher effective demand. Then, however, it is still not certain that
demand increases sufficiently to compensate for total investment cost.
Even if there were no interest cost at all, effective demand might remain
too weak to afford the purchase of a larger number of goods. The effectiveness of monetary policy in raising real investment is determined by
the situation in the real economy. This is in line with the basic theory of
endogenous money.

Demand for financial assets depends essentially on two features, that is,
investors' profit expectations and liquidity preference. Financial investors
claim a sufficient profit margin. If they recognize that effective demand is
weak and that firms are not able to raise their sales revenues sufficiently
to cover investment cost and profit requirements, they do not purchase
securities. Given that monetary policy does not improve conditions for the
real economy from the perspective of companies, neither does it in the view
of financial investors. However, even when effective demand is weak and
production is in a slump, the second feature, liquidity preference, may still
respond to changes in monetary policy. It is still valid that a lower interest
rate makes saving in a bank deposit less attractive. For banks it is easier to
refinance themselves so that they can take more risks. Professional investors can increase their leverage at a lower cost. These mechanisms may
lead to a rise in demand for financial assets even when economic prospects
are doubtful. The interest rate level serves as a kind of reference benchmark. Dividends and bond rates stagnating at a certain level become more
attractive the more the benchmark interest rate falls. Growing asset prices

triggered by a drop in liquidity preference are a signal to attract more financial investment so that prices climb further.

The magnitude of the reaction of financial asset supply and demand to changes in monetary policy can be considered as elasticities. In our case, they are the interest rate elasticities of financial asset supply and demand. While the elasticity of financial asset supply depends almost exclusively on the state of the real economy, the elasticity of financial asset demand, in addition, relies on liquidity preference. Real economic conditions change according to the business cycle. This means that supply and demand elasticities are not stable but fluctuate over time and so does the effectiveness of monetary policy. When the economy enters a boost, expectations are optimistic and effective demand grows. Monetary policy is able to support this upsurge by lowering the interest rate level. The more the economy converges to full-capacity utilization at the summit of the business cycle, the less effective monetary policy becomes, since total output cannot be larger than full-capacity output. In a downswing or outbreak of a crisis it becomes harder for the central bank to confront worsening performance. Employment decreases and effective demand weakens. A too depressed economy is hardly possible to revive only by means of monetary policy.

The case of liquidity preference is more complicated. It is affected by monetary policy as well as by the business cycle. In booms, dividends are high, risks of financial assets are judged to be low by investors and therefore their profit expectations are optimistic. Agents have low preference for liquidity. The latter grows as soon as financial markets turn down. The effect of monetary policy on liquidity preference is likely to be larger when economic conditions are stable and to be smaller when fluctuations of securities prices are high. High price increases come along with waves of optimism. Crashing prices occur together with strong pessimism and a run into safe money or government bonds. In such situations, the central bank faces troubles in counteracting these waves by bringing down or raising liquidity preference. In contrast, when financial asset prices are stable, that is, either at a low level in a recession or depression or in a time of permanent good performance, a change in the interest rate should have a stronger impact on liquidity preference.

All in all, it can be stated that the strength of the impact of monetary policy on financial asset supply and demand depends on the state of the real economy. Since interest rates are in general positively correlated with liquidity preference, expansive monetary policy tends to raise demand for financial assets by lowering the liquidity preference. Thus, besides profit prospects of financial assets, financial investment is influenced by monetary policy through a second driving force, which is effective at a stronger or weaker magnitude depending on the state of the real economy. Under these conditions, demand for financial assets, that is, financial investment, tends to react stronger to monetary policy than supply of financial assets, to wit, real investment. In the case of a cut in the interest rate, financial asset demand grows more than supply. For supply and demand to be equal, asset prices have to increase. This is the central result of this alternative approach: prices of securities can, in principle, rise even when there is no change in economic fundamentals. We arrive at a conclusion that is at odds with the efficient markets hypothesis in neoclassical theory. Given that the economy is in a slump, real investment increases only marginally in response to a lower interest rate. For investors, it becomes nevertheless more attractive to purchase financial assets in response to a fall in liquidity preference.

In this sense, the financial asset market and its relationship with monetary policy can be characterized by an exogenous constraint given by real economic conditions. Once firms have decided on their investment volume, supply of financial assets is given. From this point onwards, securities prices are determined by demand from financial investors.

Rising demand for financial assets in light of expansive monetary policy may sometimes be overshadowed by an opposite effect. Active attempts of the central bank to influence the economy are sometimes interpreted in a contrary manner by investors. For example, a sharp cut in the interest rate might be a signal that the economy needs support, that is, that its outlook is worsening. This can induce investors to sell securities and to let prices fall. In that case, a lower interest rate leads to lower demand for some financial assets (Neely, 2011, p. 23). Yet, these effects occur at the short horizon. In the longer run, prices drop more owing to actual downturns. They do not jeopardize the alternative view of financial markets but confirm it, by stating that demand and supply sides of financial assets evolve partially independently from each other.

It can be stated that the alternative view gives an explanation of financial market development and enlightens why financial asset prices are not just a simple reflection of fundamentals. While the quantity theory of money implies that an increase in the money stock leads to a proportional increase in the level of consumer prices, the concept of endogenous money allows the representation of reality more appropriately. The elasticity of money, when it is recognized as endogenous, reveals that money is not necessarily created merely for the needs of the real economy. Financial markets have an autonomous role, where money can flow without being employed in production. It is thus a logical argument that enhanced liquidity induced by expansive monetary policy gives rise to increasing prices of financial assets.

Obviously, the alternative view is derived against the background of financial market developments of the past two decades from the mid-nineties to this day. The keyword of financialization is a widely discussed issue. While it is not of great concern from a neoclassical perspective, it is investigated here for its effects on the connection between monetary policy and financial markets. A broad definition of financialization is given by Epstein (2005, p. 3), who characterizes financialization as 'the increasing role of financial motives, financial markets, financial actors and financial institutions in the operation of the domestic and international economies'. It is a rarely doubted fact that financialization has taken place in the past two to three decades. Profits and income shares of financial institutions have increased in the past (Epstein, 2005, p. 4). The importance of shadow banks has clearly augmented. Net liabilities of shadow banks multiplied by more than four from the early 1990s until the breakout of the financial crisis in 2007 (Pozsar et al., 2013, p. 7). Shadow banks are not actual banks, as they do not have the power to create money. They do not have access to central bank reserves and therefore need a private or public financial backstop (Claessens & Ratnovski, 2014, pp. 4–5). They can only collect existing savings and recycle them by providing new loans out of these savings (Sawyer, 2013, p. 233). Shadow banks have the role of a financial intermediator. As such, however, they can well exert a significant influence on financial market performance. Money creation is enhanced indirectly. By mobilizing savings that would otherwise just function as passive deposits, they raise demand for financial assets. A commercial bank that is confronted with a drain of deposits then has to demand additional central bank reserves. Since many of these non-bank financial institutions aim at highly profitable and risky investment, they employ high-leverage strategies. This requires a higher volume of bank loans that would otherwise not have come into existence. A second probable effect is that shadow banks raise the volume of bank credit indirectly by providing and marketing investment opportunities. This makes financial investment more attractive and might lead investors to riskier behaviour. Enhanced shadow banking activity may also be a symptom and thus a proof of larger credit creation. When the balance sheets of traditional banks grow too long, they run the risk of violating reserve requirements or growing short of liquidity. They intensify liquidity management by removing assets and liabilities from the banks' ledgers. The new vehicles where assets are outsourced represent a kind of shadow bank.

In practice, many shadow banks are provided with contingent lines of credit or credit put options like wraps guarantees or credit default swaps by commercial banks (Pozsar et al., 2013, p. 4). These services constitute a private backstop for shadow banks and are a reason for large credit creation in good times.

Financialization enhances the mobility of liquidity. It makes capital allocation more efficient, however in a quite limited sense. Changes in monetary policy are therefore faster and more rigorously exploited. Together with highly leveraged investment strategies, overreactions of financial markets to changes in the interest rate are more likely. The long-run process of financialization in the past decades implies that the results of the present alternative view of financial markets have got more weight. The relationship between monetary policy and financial markets should therefore have become tighter over time.[11] This should be kept in mind for the remainder of our analysis.

The findings of this chapter are kept in a synthetic manner to stress its main features. It yields a pattern that draws a simplified but basically adequate picture of reality. For instance, the relatively strongly growing US economy of the noughties of the twenty-first century was marked by a high and still growing degree of financialization, well-performing financial markets as well as rising inequality of incomes (see Tomaskovic-Devey & Lin, 2011). The latter was a threat for economic growth, since falling incomes in income classes, where the propensity to consume is high, drag down effective demand. To avoid a drop in consumption, lower and middle classes started getting indebted (Stockhammer, 2012, pp. 14–15). The best-known result of this was a market full of subprime mortgages, a topic that has been discussed in abundance. Low income and growing debt on one side of the social spectrum was growing profit on the other side. Those profits were reinvested by purchasing financial assets. Hence, inequality was a source of increasing money creation that was in the end not used to purchase real goods but financial assets. Monetary policy of that period kept interest rates at a permanently low level. While this ought not to be the only, or principal, source of rising asset prices, as claimed for instance by representatives of the Austrian school (see for example, Murphy, 2008), it was at least a contributing factor. Inequality and indebtedness were only allowed to last so long and to climb so high because purchasing power in the form of consumer credits and mortgages was accessible at easy conditions. The economic crisis, at least temporarily, put an end to these trends.

In this chapter we have explained how monetary policy can lead financial markets away from their fundamental values. Central bank actions can distort the relationship in between. Since there is no reason to assume a certain equilibrium path, monetary policy impacts do not necessarily fade out fast. Securities prices can rise without there being necessarily a rationale in the real forces of the economy. However, high profits in financial markets cannot be realized forever, if the real economy's long-run growth potential is stagnating or only slowly increasing only. Sooner or later, there will be a lack of effective demand, and production profits drop. At this point, the neoclassical explanation would refer to the efficient markets hypothesis, which guarantees that asset prices always adapt efficiently to changes in fundamentals. Fluctuations in fundamentals, that is, in the business cycle, are treated best if they are not counteracted (Goodfriend, 2007, p. 26). An economy without nominal rigidities reacts to exogenous shocks in the most efficient way. For the current crisis breaking out in 2008, neoclassical proponents recommend, likewise, that public interventions generally worsen the situation by distorting otherwise efficient resource allocation (see, for instance, Fernández de Córdoba & Kehoe, 2009, pp. 6–7). The alternative view suggests a less harmonic mechanism. The periodic occurrence of financial crises reveals a contradiction between financial markets and real forces. It can be traced back to the basic contradiction between effective demand and profit acquisition that is inherent to capitalist economies. This approach allows for volatile and discontinuous economic processes where distortions and disruptions are endogenous features rather than exogenous shocks.[12] Policy interventions may be required since the economy is possibly not able to find its way out of these contradictions. This will matter in the remainder of this research work, as we know that past decades were characterized by discontinuities that need also to be explained in the investigation of the oil price evolution.

2.2 MONETARY POLICY EFFECTS ON THE MARKET FOR CRUDE OIL

The general analysis of money and financial markets provides a consistent explanation and a useful basis for our further research. To apply the findings to the crude oil market, some extensions and modifications have to be made. Yet, we keep the focus on the monetary mechanisms of the economy, as has been done before, as it allows for detailed insights into the building of oil prices.

2.2.1 Crude Oil as a Commodity and a Financial Asset

The investigation of the oil market reveals that the economics of oil are specific. As has been noted by several authors (see, for instance, Fattouh, 2010, p. 13), crude oil has a double nature. On the one hand, it is a common commodity among others, serving as a raw material mainly in energy production. There is a world market where supply and demand, that is, production and consumption of oil, meet. On the other hand, oil is the underlying object of a financial asset that is traded in commodity futures exchanges and over the counter (OTC). Futures are standardized contracts that refer to the exchange of predetermined units of an object at a predetermined price on a predetermined date in the future where account settlement takes place in a clearing house (Volkart, 2008, pp. 938–942). In the case of crude oil, the producer commits to the delivery of a unit of oil on the said date while the consumer commits to purchase it. The delivery position is called 'short', the purchasing position is named 'long'. In the practice of complex financial markets, however, there can be many intermediate steps such that it is neither necessarily the original producer on the short side nor the final consumer on the long side. Moreover, a holder of, say, a long position may buy an additional short position to offset her liabilities. In that case, no delivery takes place. As a general notation, short positions refer to the supply side while long positions represent the demand side of futures contracts.

Owing to the dual nature of crude oil (both a commodity and a financial asset), there exist two markets for one and the same good. In the spot market where oil is a commodity, it is exchanged in physical quantities. It is embedded in the real-world economy and thus influenced by industrial production, consumers' income, geopolitical conflicts and wars, market structure and new oil discoveries. As additional factors, there are considerations with regard to time. Oil is an exhaustible, fossil resource. All participants in the market have conjectures about global oil reserves and therefore take partial precautions (see Kilian, 2009b), for example, they may build oil inventories. Moreover, technological progress may give rise to changes in industrial production, to a reduction in the use of motor fuels or economization of house heating. Trends in the opposite direction may materialize as well (see, for instance, Anger & Barker, 2015). In the long run, the oil intensity of economic output can change.

In the futures market for crude oil, oil is traded in the form of contract papers rather than physically. Someday, however, a futures contract eventually has to be fulfilled by the delivery of the agreed quantity of oil. The futures market can, in principle, be divided into several markets depending on the time span of contracts. In general, at all horizons, the crude oil futures market has some distinct features. Paper trade is more flexible than trade of physical oil. The futures market reacts therefore faster to new information. This is why trading activity is likely to be more volatile. Considering futures as financial assets means that they are in the focus of financial investors who do neither have an interest in the consumption of oil nor in its physical trade. The market for crude oil futures is in many aspects analogous to financial markets analyzed in the preceding chapter. Yet, some additional features of the futures market need to be taken into account.

The futures market reflects the future of the crude oil market. As such, however, it is already currently present. It exists next to the spot market where oil is physically traded. The futures market of today is the spot market of tomorrow and the spot market of today was a futures market yesterday. Spot and futures markets cannot be separated from each other. Yet, given a specific point in time, there are always both a spot and a futures market. Hence, they cannot be the same. As will be seen, a futures market is different today compared to tomorrow when it will be a spot market. Oil producers sell oil and issue financial assets in the form of futures contracts. Financial assets other than futures are issued by corporations on the one hand while they sell their produced goods on the other. In the case of oil futures, the produced good for sale and the financial asset is one and the same object. Correspondingly, investors consume or at least trade oil and purchase financial assets at one and the same time. They often do these two things with different motives but they rely on the same object, which is crude oil.

In analogy to the preceding chapter, the spot market reflects the fundamentals of the oil market. The futures market, on the other hand, has financial market features. Referring to the efficient markets hypothesis, neoclassical approaches suggest that the futures market is a simple reflection of the spot market. Changes in oil market fundamentals cause changes in futures prices but trading activities in the futures market do not influence the spot price of crude oil. They help producers and consumers to hedge their future production and consumption at a specific price in order to avoid the risk of price fluctuations (Fattouh et al., 2012, p. 15). Speculation in the futures market, according to neoclassical theory, is helpful and necessary to support efficient price discovery and risk transfer (ibid., p. 7). In contrast to this view, we will argue that the futures market has an autonomous role. It is not independent from fundamentals but there exist complex two-sided interrelations between futures and spot markets.

Since futures short positions are, unlike stocks, not only liabilities to buyers but represent also the produced good in the same object, supply of futures is also supply of crude oil. Higher prices give firms an incentive to increase oil supply either in the form of real oil deliveries or by contracting new futures. Total supply of physical crude oil is constrained by present and future production capacities. The amount of oil ready for delivery within a horizon of, say, up to one year, finds its upper limit when all oil produced at full-capacity utilization and all inventories built in the past are either sold physically or securitized. In contrast, the supply of futures contracts, to wit, the number of short positions, is basically unconstrained. It does not make sense from a macroeconomic point of view to contract the delivery of 80 million barrels of crude oil on a particular day in the future if only 70 million barrels are available. But from the microeconomic perspective of a single oil company or an investor, going short when prices are high is profitable. If they want to avoid physical delivery, they have to offset their short position by the purchase of a long position.

On the demand side, purchase of physical oil is constrained by demand in the real economy. Yet, demand for futures is potentially unlimited. Investors can, in principle, raise their long positions infinitely. In contrast to stock markets, demand for oil futures does not depend on profit expectations of firms because there is no dividend. But likewise, investors build expectations about price changes. Another motive is again liquidity preference. Lower liquidity preference means *ceteris paribus* higher demand for futures. Hence, in analogy to our financial market analysis, demand for futures can increase without having a preceding change in fundamentals, that is, the oil spot market and the rest of the economy.

The number of futures contracts can grow infinitely since every position can be evened up by a new counterbalancing contract either by an oil company, a gasoline producer, a financial investor or whoever. The simple formula (2.1) still holds. Demand for financial assets is always equal to supply of financial assets; long positions are always equal to short positions since all futures contracts require two contracting parties. As a difference to conventional financial assets, supply and demand sides of the futures market are more flexible. This means that unlike stock or bond supply, the supply of futures is not constrained by real forces. Given a higher demand level and thus a higher price of futures, supply can increase without limit such that the number of contracts is basically allowed to rise to a higher and higher level. It is, however, wrong to claim that, since long and short positions are always equal, supply and demand behave necessarily equally. This would imply a permanently stable price. The logical equality of long and short positions does not say anything about underlying market developments (see, for instance, Irwin et al., 2009, p. 379). If a sufficient number of investors expect a higher spot price in the future, they raise their demand for long positions. By purchasing a futures contract at a given price, they hope to sell it at a higher price before the delivery date or to receive delivery of a certain oil quantity that has a higher spot market price than the contracted price they have to pay for it. Prices offered by demanders increase with the number of investors in order to get matched with a supplier. Futures suppliers receive the signal of a higher price and therefore have an incentive to raise their short positions. The equality of supply and demand is not distorted in any given moment. In the other direction, if a sufficient number of agents expect a decreasing spot price, they raise their short positions. If the price effectively decreases, they can deliver crude oil at a higher contracted price than the actual market price. Thus, depending on whether a change enters the futures market either on the supply or the demand side, the price reacts in a different way. On the demand side, there is the feature of liquidity preference in addition to pure price expectations. It can as well exert an influence on price and is emphasized when introducing monetary policy to the futures market.

These reflections reveal that participants in the futures market must be heterogeneous. For two parties to contract on the long and short side they must diverge in their expectations of future price evolution, in their risk aversion as well as investment strategy. Once this is recognized, it becomes harder to maintain the efficient markets hypothesis. Given that investors pursue counteracting investment strategies, a wide field of possible price and quantity outcomes opens. Of course, investors build their expectations based on developments in the real economy. Finally, all oil that is produced must be sold physically. Oil can only have a price if it is of practical use. The need for oil is essential for futures contracts to make any sense from a macroeconomic point of view. To agree on a contract on the long position without having an interest in the consumption of oil or being able to sell it to someone else is certain to be a losing deal. Total sales of oil are constrained by effective demand in the spot market. But uncertainty in the futures market, and the large elasticity of futures markets in their number of contracts as well as contracted prices, allow for a high variety of evolutions that can be closer to or further from fundamentals. This is to be investigated in more detail.

The connection between the futures market and the spot market

The basic mechanism that connects the futures market to the spot market is shown in Figure 2.1. Yet, representing a financial market and especially a contract market in this way is not without its problems. The futures market in the left-hand panel exhibits both rising supply and demand curves. This can be justified by the above argument that suppliers and demanders of futures send and receive price signals. A higher futures price makes it more attractive for suppliers to contract future oil deliveries. Investors may on the other hand expect a rising price and react by increasing their long positions. The causality may also be the other way around such that a higher futures price gives an incentive to raise long positions. But, in principle, reality might also prove different. If behaviour on the supply side is speculative, an expected price drop will raise short positions such that there is a negative correlation between price and the number of short positions. Thus, the structure of the futures market – like any market – depends on the motives of market participants and their expectations (see, for instance, Pilkington, 2013, pp. 13–22). Past performance may as well impact on investors' behaviour as an additional factor. All these features change over time. The oil futures market can therefore be considered as a typical case of radical indeterminacy in capitalist economies (see, for example, Varoufakis et al., 2011, pp. 294–298). For the purpose at hand, the proposed modelling seems to be the most appropriate one. In contrast, it is conclusive to assume that the oil spot market is in a large number of cases composed of a rising (or vertical) supply curve and a declining demand curve. The conventional feedback mechanisms between oil price, supply and demand are at work. A higher price lowers demand, raises supply and so on. But even in the spot market, expectations may change the slopes of curves in the short run. So both in spot and futures markets, feedback mechanisms can be the conventional ones but they can also go in the opposite direction depending on agents' expectations. Our presentation is a specific choice out of many alternatives.

Another problem arises with regard to the issue of time. As explained, the futures and spot markets take place at the same time but refer to a different time period. The futures market in the left-hand panel of today will become the spot market in the right-hand panel of tomorrow. Our consideration of the relationship between them must therefore take place in a single point in time, that is, in a single production period. Supply and demand curves in the diagrams cannot be long-run curves since, as such, the spot market would include present as well as future behaviour of suppliers and demanders. For example, to assume a horizontal supply curve in the spot market as an expression of oil producers' ability to fulfil all

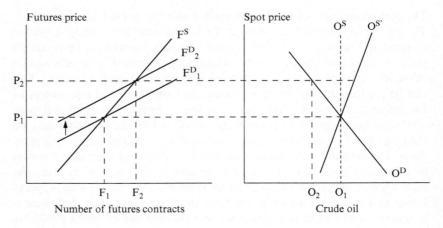

Source: Author's elaboration.

Figure 2.1   Connection between futures and spot markets

demand in the long run, would imply that the spot market in the future, that is, the futures market today, is contained in the spot market diagram. This would confuse our analysis since we want to map the future spot market separately in the left-hand diagram. The supply curve in the spot market hence is a vertical line, O^S, because production is fixed in a given period. Assuming that firms supply existing inventories when the price is high and accumulate additional inventories when the price is low turns the vertical line to an increasing supply curve, $O^{S'}$. The problem of this particular case reveals the general failure of the neoclassical equilibrium approach to deal with real time. These diagrams do not have a time dimension and thus are not able to illustrate how equilibria are reached.[13]

For our purpose, however, Figure 2.1 is a tool to show a particular mechanism rather than a complete theoretical model. The general equilibrium view has already been criticized and rejected in the previous chapter. In spite of almost absolute uncertainty, the diagram can help to enlighten the financial market impact in the oil market. Assume a demand increase in the futures market such that the demand curve shifts from F^D_1 to F^D_2. Where such a shift may come from is of secondary interest now and will be investigated in relation to monetary policy. Higher demand for financial assets will raise the price from P_1 to P_2 but also enhance short positions. The demand shift translates into both a price effect and a quantity effect in the form of larger open interest. Assuming away real complications, the futures price and the spot price are equal. Deviations create arbitrage opportunities that are exploited. In the current case, the higher price in the futures market gives producers an incentive to raise supply in the futures market. All oil that is sold in the spot market at a price lower than P_2 is a losing deal for producers, as they have the opportunity to offer it favourably in the futures market. Demand in the spot market decreases, therefore, but is offset by higher demand in the futures market. In the face of higher demand in the futures market and lower demand in the spot market, companies adapt respective supplies of futures and physical oil just to maximize expected profits. Hence, supply and demand sides in both futures and spot markets behave such that futures and spot prices are equalized.

In reality, futures and spot prices are not necessarily always exactly equal. As outlined by Kaldor (1939, p. 6), they usually differ structurally. The variables that make the difference are the interest rate level, the risk premium, the convenience yield and the carrying cost of the commodity. Kaldor (1939) shows that under normal circumstances, that is, in a non-speculative environment, the expected futures spot price is equal to the current spot price since expectations about the future are transferred to the present, as shown in Figure 2.1. The futures price is lower than the expected price by the amount of a certain risk premium. The risk premium is the compensation of the risk of future price fluctuations that moves from the seller to the buyer when the uncertain futures spot price is fixed by the current futures price. The futures price also differs from the current spot price. The difference is made up by the interest rate level and the net convenience yield. The purchase of a futures contract in contrast to a purchase in the spot market has the advantage that the purchaser only has to pay at the delivery date in the future. For the remaining time, the investor can invest the capital at the current rate of interest. On the other side of the contract, the seller does without immediate payment and therefore loses potential investment returns at the current interest level. The net convenience yield, that is, the convenience of having a commodity physically available at any moment minus the cost of storing it, lies on the seller's side because the commodity has to be delivered only in the future rather than today. The buyer on the other hand does without the net convenience yield until the delivery date. The futures price is therefore determined by the spot price plus the return of an alternative investment at the current interest rate minus the net convenience yield. At this point it can easily be shown that since the risk premium is always positive, the futures price is always lower than the expected price and therefore lower than the spot price in a non-speculative environment. Hence, the convenience yield is larger than the interest rate and the futures price is lower than the current price. This is what Keynes (1930b/2011, p. 143) calls 'normal backwardation' as the futures price is usually lower than the spot price under normal conditions.

However, nowadays the case is less clear. Given that some financial investors do not have any interest in the physical possession of oil, they do not require compensation for escaped convenience yield. It is not just that they purchase futures contracts, which of course existed already when Kaldor wrote his paper in 1939, it is that, due to the larger variety of present-day financial actors compared with those in the past, many of them consider futures not as a symbol of a physical commodity but simply as a financial asset. As such, futures are completely disconnected from the physical properties of oil when traded by financial investors. The difference between the futures price and the spot price then is only made up of the positive return of interest. The futures price is thus higher than the spot price; the market is in 'contango'. Additional features complicate the situation. The spot price may well be higher in a situation where there is an acute delivery bottleneck that will be relaxed in the future. Alquist and Kilian (2010, p. 562) present a model where the convenience yield varies over time depending on expectations about future supply disruptions. When uncertainty about the future is pronounced, the convenience yields increases such that demand in the spot market increases relative to the futures market. The spot price increases relative to the futures price. The model seems to fit data remarkably well for a part of the time frame considered but fails for the rest of the observation period (ibid., p. 566). Even though correlation values are rather impressive, the model does not uncover all possible causalities. The existence of financial investors, who consider oil exclusively as a financial asset, is suppressed.

Whether one observes 'contango' or 'backwardation' in the oil market depends on various factors. There are the traditional influencing variables remarked by Kaldor (1939). Furthermore, the motives and expectations of market participants play a role. Time delays in the spot market may be the source of additional differences. The difference between the futures price and the spot price does not allow drawing any premature general conclusions about causality. Neither does 'contango' necessarily imply an influence of the futures market on the spot market nor does 'backwardation' state the opposite to be true. The price structure is undetermined and can take various forms depending on the situation. In fact, crude oil futures and spot prices used to move together quite closely, as data show. While the market tended to be in 'contango' around the price peak of 2008, 'backwardation' dominated in the years before (EIA, 2015c).

Several studies empirically estimate the relationships between the crude oil futures and spot markets (see Kaufmann & Ullman, 2009; Silvapulle & Moosa, 1999). The general result is bidirectional causality between the futures price and the spot price. Changes in the futures price impact on the spot price as well as the other way around. Many empirical studies of this kind test for Granger causality, that is, whether a change in one variable helps predict a change in another variable in a statistically significant way (see, for example, Granger, 2004, p. 425). Such a definition of causation is quite generous and leaves ample room for further analysis since it is not immediately clear which forces effectively drive one price to influence the other. Moreover, especially in the presence of bidirectional causality, dynamic interdependencies between the futures and spot markets arise.

For the remainder, it seems to be more productive to accept the existence of numerous simultaneous factors that impact on the connection between the futures market and the spot market without trying to quantify and assess each of them. Deriving essential results from the spread between the futures price and the spot price and relying on them in further analysis seems to be a rather unstable way of proceeding. The price spread changes over time and so do its driving forces in a complex way. What can be concluded from a logical point of view and from empirical data is that the futures price and the spot price follow a common pattern owing to arbitrage and mutual influences between both markets. These tight connections are especially due to the nature of oil being both a commodity and a financial asset. In contrast to a large part of existing literature, we choose an approach that goes in the opposite direction. We do not start with the analysis of the difference between spot and futures prices since they are not crucial for general price development. Rather, we start by taking their equality on analytical grounds, then allowing for distortions.

The role of inventories
In an influential paper, Hamilton (2009, pp. 234–240) shows, by means of a simple model, that the effect of a demand shock in the futures market on the spot price depends on the price elasticity of gasoline demand. The interpretation is analogous when referring to the demand elasticity of crude oil and is also reflected in Figure 2.1. If the demand elasticity is zero, the demand curve is a vertical line. Oil consumption is constant at all price levels. A price increase originating in the futures market as simulated in Figure 2.1 does not then change the quantities in the spot market. Since oil production of that period is also given by a vertical line, O^S, inventories do not change. The price can, in principle, grow to infinity without any reaction on the demand side of the spot market. In fact, demand elasticity is rather low. Most estimates range from between –0.005 and –0.02 (Krichene, 2002, pp. 568, 570) to values of about –0.1 (Cooper, 2003, p. 4) depending on the time horizon of consideration. Kilian & Murphy (2014, p. 474) get a larger estimate of –0.24. Despite probable challenges for each estimation method, elasticity of demand is different from zero. The demand curve in the right-hand panel in Figure 2.1, O^D, may be close to a vertical line but still has a negative slope. Thus, all price increases lead to a fall in spot market demand. While the situation may still be profitable for producing firms as they can sell oil in the futures market, real oil quantities sold in the spot market decrease.

As Hamilton (2009, p. 238) argues, for the spot price to stay at the level induced by higher demand in the futures market, oil inventories have to be accumulated. Following this logic, it is only with growing stocks that there is sufficient scarcity in the market to keep the spot price high. In Figure 2.1, inventories have to increase by the distance between the continuous demand curve, O^D, and the dotted vertical line of oil production, O^S, at the level of the upper dashed line, P_2. The larger the price elasticity of demand, the larger is the required quantity of oil to accumulate in stocks in order to keep the oil price at a given level. This finding leads Hamilton to the conclusion that too large an elasticity of demand makes any potential speculative influences from the futures market to the spot price insignificant (ibid., p. 238). All futures price increases induced by financial investors generate a strong reaction of demand in the spot market. Investors then recognize that their price betting is irrational and does not correspond to fundamentals. They stop purchasing long positions since this behaviour is certain to yield them financial losses.

There are good reasons for criticizing this view, which is based on the efficient markets hypothesis. It is true that oil producers eventually rely on effective demand in the spot market. Financial investors know that. However, due to uncertainty, speculating on an increasing oil price without there being changes in the real economy can nevertheless be rational from the perspective of a single agent. In this respect, the futures market works analogously to stock markets. But there is an additional feature that supports the effectiveness of speculation. It is due to the dual nature of crude oil as a commodity and a financial asset. It is obvious from a macroeconomic point of view that effective demand is rooted in the spot market. From the microeconomic perspective of an oil company, in contrast, effective demand occurs in the futures market as well as in the spot market. Higher demand for futures brings as much profit for companies like selling oil at the same price (or small differences due to 'contango' or 'backwardation') in the spot market. Hence, as a crucial conclusion, companies do not mind accumulating oil inventories. In their view, these inventories are not reserves. They are already sold. A high elasticity of demand is therefore not *per se* a barrier to the effectiveness of speculation.

In the previous chapter we pointed out that contradictions between financial markets and the real economy cannot be carried on without ending in disruption. The same applies for the market of crude oil. Inventories cannot grow forever, as we will show in more detail below. But individual investors do not share the macroeconomic perspective. They face uncertainty and cannot escape competition. An oil company that continues supplying oil in the spot market at a lower price without benefitting from increased demand for futures behaves irrationally. It is again the fallacy of composition stating that what is rational from a microeconomic perspective is not necessarily rational from a macroeconomic one.

Let us suggest as a counterargument that if speculative influences should be significant in the spot market, a corresponding increase in oil stocks must necessarily be observed. Any failure to assess rising inventories in times when the oil price increases means that speculation cannot be the source of the rising price (see, for instance, Alquist & Gervais, 2011, pp. 8–9; Krugman, 2008). However, this empirical exercise is impeded by several difficulties. First, the acquisition of inventory data suffers from the lack of a clear definition of inventories and from political influences. Instead of building inventories, firms may also leave more oil in the ground, which may help them save on carrying costs. Both are, in some sense, inventories. Official inventories include produced oil and petroleum products but ignore underground 'stocks' (EIA, 2015d). The case of capacity utilization is similar to this argument: in neoclassical theory, firms usually produce at full-capacity utilization.[14] Implicitly, Hamilton (2009) accepts this claim. It is only in this case or in the case where producers have constant spare capacities that a speculative price increase leads to higher inventories (given that demand elasticity is negative). If oil producers vary the degree of capacity utilization while speculation in the futures market affects the oil price, the inventories can basically reach any volume and do not necessarily have to increase.

Second, certain parts of the inventories are made up of strategic reserves. The political nature of these reserves implies that their amount is not necessarily transparent. Moreover, oil stock data used to be merely about OECD countries and thus exclude other important countries like the BRICS.[15] Kilian and Lee (2014) employ an alternative broader estimate of inventories. However, it faces potentially reduced but similar concerns.

Third, uncertainty can, in principle, lead to paradoxical results. If accelerating demand by financial investors for futures raises the oil price, all market participants start building expectations. If oil consumers extrapolate past development to the future, they may increase demand in the spot market today even though the price is higher than it was yesterday. This happening occurs when the demand curve in the spot market temporarily takes a positive slope. Such a mechanism is not likely to be of long duration since demand is constrained and cannot react to all possible expectations. However, inventories might then decrease, albeit the source of the price boost is of a speculative nature. Kilian (2009b, p. 1059) defines a shock that reflects precautionary demand by consumers in the face of an uncertain future. However, such precautionary demand may be only a response to foregoing the speculative activities in the futures market that are responsible for the price change. Thus, it is not only the existence of precautionary demand that should be taken into account but also its connection to speculation.

Finally, Hamilton's (2009) view is incomplete. A price elasticity of demand of zero or close to zero is neither a necessary nor a sufficient condition for speculation to be effective. It is not necessary because it can take place with a larger range of elasticity values. On the other hand, the condition is not sufficient because it only takes the demand side into account. However, it is not only demand but also supply that reacts to a higher oil price. While demand falls to a larger or smaller extent, supply is likely to rise owing to investment in additional production capacities. This aspect is quite important when investigating long-run speculative impacts on oil spot market quantities.

A monetary analysis of the futures market

Let us now emphasize the interdependencies between the futures market and the spot market for crude oil in a monetary analysis. It will be seen that the endogeneity of money plays a crucial role. Table 2.4 displays two balance sheets of an oil-producing company and a financial investor, respectively. For the simplification of accounting, other market participants are excluded in this particular examination. Complications are added after but do not change the basic results. In the initial point, (1), the oil company holds x barrels of oil as inventories. For further simplicity, we ignore other balance-sheet values like equipment, real estate and the like. Moreover, the company is fully self-financed by assumption. This means that all liabilities consist of equity E. The equivalence of assets and liabilities implies that oil inventories valued at their current market price are equal to equity, that is, $x^*p_1 = E$. Now, the company goes short in the futures market and contracts on the delivery of crude oil in the amount of current inventories x. The contracted price is p_1 in order to keep it simple. On its assets side, the well-known and certain return of future oil delivery is anticipated and considered as a balance-sheet value. One may discount these future returns by some interest rate but it is needless for this analysis. On the liabilities side, there is the equally well-known future oil delivery commitment, which is in fact a kind of debt.

The other side of the futures contract is met by a financial investor. She goes long by the same quantity x valued by p_1. It is an asset since it provides her with a real good in the future, to wit, crude oil. On the liabilities side, there is debt because the investor owes a price p_1 for each barrel of oil to the company. If many financial investors purchase futures long positions, demand pressure raises the futures price and the spot price to a level of $p_2 > p_1$. The investor does not have an interest in real oil delivery. Therefore, she has to offset the long position before the expiry date of the contract. A short position of equal size is purchased. The financial investors' balance sheet lengthens by the new book-entry (2). Future sales earnings are added to the assets while the commitment to future oil delivery enters on the liabilities side. The offsetting short position has been sold by the oil company. Under (2) in the balance sheet of the latter, the deliberation from the future delivery commitment enters the assets side. This balance-sheet item is greater than the short position earlier on, because oil is now priced at p_2. On the liabilities side, missed future earnings from oil delivery enter. Crossing out the short position and its annulment yields an increase on the assets side by $x^*(p_2 - p_1)$. This is due to the higher pricing of crude oil inventories, which remain stocked rather than being sold. On the other hand, crossing out future sales earnings and their subsequent cancellation yields a debt of $x^*(p_2 - p_1)$. Under (1), the oil company had a balance at the futures market clearing house of x^*p_1 due to its delivery commitment. Under (2), the selling of the short position at the new market price means that the company pays x^*p_2 to the purchaser of the short position for being deliberated from a future oil delivery. Thus, it remains an obligation of $x^*(p_2 - p_1)$ at the clearing house. As can be seen under (3), clearing-house debt is offset by crude oil inventories on the assets side of which each barrel of oil is now valued at a higher price.

For the financial investor, crossing out future sales earnings and initial debt as well as the long position and the offsetting short position brings a clearing-house balance of $x^*(p_2 - p_1)$ in the form of cash in (3). It represents a speculative profit. In this example, offsetting occurs by the purchase of the company's short position. The financial investors and the company also might, in an analogous way, close a new contract where the financial investor goes short and the company goes long. The result would be the same.

Table 2.4 explains in the framework of Figure 2.1 how speculation can impact the price of oil to the benefit of the speculator. On the other hand, speculation is not to the damage of the oil company. In our example, losses in the futures market are compensated by the higher price of oil

Table 2.4   Balance sheets of an oil company and a financial investor

Oil-producing company

	Assets		Liabilities	
(1)	Crude oil inventories	x^*p_1	Equity	E
	Future sales earnings	x^*p_1	Short position: future oil delivery	x^*p_1
(2)	Selling short position: no future oil delivery	x^*p_2	No future sales earnings	x^*p_2
(3)	Crude oil inventories	x^*p_2	Equity	E
			Debt	$x^*(p_2-p_1)$

Financial investor

	Assets		Liabilities	
(1)	Long position: claim on oil in the future	x^*p_1	Debt	x^*p_1
(2)	Future sales earnings	x^*p_2	Short position: future oil delivery	x^*p_2
(3)	Cash	$x^*(p_2-p_1)$	Profit	$x^*(p_2-p_1)$

Source: Author's elaboration.

inventories. Hence, the higher oil price does not directly result in larger cash holding but inventories can be sold in the global liquid crude oil market. The price increase suffices exactly to pay back the debt. The oil price is now at a higher level without a single barrel of oil having moved anywhere, that is, without any change in fundamentals.

Some extensions reveal the mechanisms even more drastically. Assume that the oil company offers only a part of its inventories in the form of futures contracts or, similarly, only contracts for a fraction of future oil production. All else equal, the price increase again involves a loss for the company's short position in the futures market clearing house. But now, there is a benefit given by the higher oil price multiplied by the total stock of inventories and future oil production. It is not only contracted oil barrels that are valued at a higher price but all existing inventories and all future oil production. This is revenue that is larger than the loss in the futures market. Thus, the company has a net profit even though it is once more in the form of oil inventories rather than realized cash.

As another extension, assume that financial investors' expectations are very optimistic. They expect a higher future spot price such that they accept a present futures price that is higher than the current spot price. One barrel of crude oil at a spot price of, say, 50 US dollars is purchased in a futures contract for, say, 75 US dollars. Recalling Figure 2.1 it is clear that both prices have to equalize apart from potential 'contango' and 'backwardation' situations. Further demand pressure in the futures market drives the oil price even higher. At the delivery date, the spot price is, say, 100 US dollars. The offsetting of the contract brings a clearing-house loss of 75–100 = −25 US dollars for the oil company and a profit of 100–75 = 25 US dollars for the financial investor. However, the new valuation of the company's inventories brings a non-realized gross profit of 100–50 = 50 US dollars. The company's net profit is 25 US dollars. The effect of financial speculation therefore brings benefits to both long and short positions of futures contracts. Contract relations in the futures market are more complex in reality. There are companies that do not only go short and financial investors who do not only go long. But since long and short positions eventually offset one another, remaining net effects are those outlined in Table 2.4.

Neoclassical economists certainly would appeal against these findings. According to their view, supply and demand curves allow for one possible equilibrium price. Deviations from equilibrium will therefore evoke reacting forces that lead back to equilibrium at a certain speed and within a certain time span (see Stadler, 1994). Our alternative analysis, in contrast, shows that market outcomes can, in principle, be placed anywhere in the quantity–price space. The true reference point is not an unknown hypothetical equilibrium but only the equivalence of financial asset supply and demand, to wit, the equivalence of short and long positions. The formula (2.1) in this chapter constrains oil futures only in the quite logical way that both sides have to be equal. When a price change in the futures market transmits to the spot market, the equivalence formula for the spot market is not violated, either. A change in the price alters demand in the spot market while production is given in the period of consideration. However, demand is still equal to production and the change in inventories. Inventories are not a market imperfection that drives a wedge between supply and demand. They are a logical outcome in a capitalist economy where there exist financial markets and where market participants build future expectations. At a given price, oil companies decide not to supply inventories to consumers but to financial investors in the form of futures contracts to which inventories are the underlying real asset. Thus, spot market demand is equal to spot market supply.

The functioning of the futures market shows quite plainly that money is endogenous. The clearing system allows investors to accumulate their open interest without limit. Opposite positions are offset by double-entry bookkeeping. Monetary payments are only required at the date of delivery or expiration of the contracts. At this point in time, existing open interest that is not offset by an opposite position is settled by delivery. In the case where futures are evened up, the financial investor gets the resulting profit or owes a loss. This is the only effective payment taking place in the course of futures transactions. In our example, the investor would be paid its profit of $x^*(p_2 - p_1)$ while the oil company has to pay a debt of the same amount. The only real capital that the investor has to invest at the beginning is the initial margin that serves as a guarantee that final payment obligations are fulfilled. If these obligations increase during the term of the futures, a margin call is carried out and investors have to raise their initial margin (Volkart, 2008, p. 941).

In the stock market the total amount of investment involves much more capital. One cannot get stocks for free and receive the return in the form of the price increase at the end. An investor has to purchase a financial asset at the current price and hopes that she will benefit from a growing price. The profit rate is given by the ratio of the price change to invested capital, that is, to the initial price of the asset (ignoring dividends). In the same way, betting on an increasing oil price requires the purchase and resale of oil even though only in the form of paper. However, a financial investor who goes long in a crude oil futures contract for one unit of oil does not pay anything on the day of contracting except the initial margin. The bill will only be met on the expiration date. Evening up the long position will not require any payment at all except in the case of a fallen spot price of oil in the meantime, which gives rise to a negative investment return. But even so, over the duration of the contract the investor is in possession of a futures position that has the power to purchase a unit of oil at the pre-determined future price. The futures contract is counterbalanced by debt at the same level, shown in the financial investors' balance sheet in Table 2.4 under (1). The other side of the coin is the oil company's future sale balance in (1) that is in turn outweighed by its short position. The analogy to the process of money creation is obvious: a futures contract creates a claim on a real good in the form of a unit of crude oil. Thus, it is endowed with purchasing power. And like money created by loans, this claim reflects an equal amount of debt. The oil company's future sale balance is a kind of loan while its short position is the balance of the investor and thus reflects a kind of deposit.

In practice, however, futures cannot be considered as a means of final payment since they are not accepted as a currency. But during the contract term, increased purchasing power in the form of futures contracts is effective in influencing investment. Paying only the initial margin rather than the total futures price allows investors to pursue speculative profits with low investment requirements. The clearing-house system makes it possible to trade paper oil with a relatively small amount of capital. Futures trading is therefore highly leveraged. The profit rate is given by the ratio of the change of the oil price to the initial margin instead of the whole futures price. Commodity speculation is therefore quite risky as it can yield high returns as well as large losses (see, for example, King, 2014).

This is basically not surprising because leveraged investments are a widely observed phenomenon in financial markets. They are usually debated from a microeconomic perspective. When a single financial investor pursues a risky investment strategy, she borrows by funding in the capital market or by taking credit from a bank. The leverage is financed by other investment entities. One investor's debt is another investor's balance. A loan is held as a deposit by somebody else in official money form. The debtor pays interest for her borrowed amount of capital. If she is not considered creditworthy, she will not be granted a loan. Interest payment and creditworthiness are factors that restrain the issuance of credit by some degree.

In the futures market, however, the leverage is a microeconomic as well as a macroeconomic one. By contracting for a future oil delivery, new debt and a new claim on real output, that is, crude oil, are created. The debt does not pay any interest and the required creditworthiness is not different from the creditworthiness of a conventional oil purchaser. The number of futures contracts is independent of the underlying quantity of crude oil available for delivery. This form of money, that is, open interest, can therefore grow to infinity as long as investors are willing to bear the risk that futures contracts contain. The macroeconomic relevance is that it is not only the stock of official money in the economy that grows. Even more than that, new capital, an unofficial form of money, is created out of nothing by futures contract conclusions. It exists outside of the official payment system and is added to conventional money as the official means of payment. The underlying debt represents a leverage that is fully unregulated since it does not bear any interest. The only connection between leveraged capital and conventional capital is the initial margin that is required to access the futures market. This shows that money – in its official form as the means of final payment or in any form – is purely endogenous and highly elastic. Money requirements of financial investors are satisfied by credit creation, whether this credit creation takes place through bank credit or through other ways.

As with financial markets in general, futures markets cannot permanently evolve independently of developments in the real economy. The price of oil futures is not at all exclusively determined by fundamental forces. There are many important additional factors from the financial sphere that impact on futures. But in the long run, all oil that is produced must be consumed in order to have a positive price. Let us first consider the demand side. Total demand for futures must grow in order to have a longer-lasting speculative effect on the price. A fraction of total open interest on the long side is held by non-speculative consumers. Their demand decreases sooner or later if the price continues to rise and so does their long position. The growth in speculative futures demand has to be larger than the decrease in non-speculative futures demand in order for the price to rise. This means that the demand shift, illustrated in Figure 2.1 by highlighting a single point in time, has to occur many times. Changing to the supply perspective, a higher oil price is likely to raise supply by means of higher inventories from past periods and investment that raises production capacities. The oil supply curve thus shifts to the right from period to period. Oil companies, therefore, raise supply of futures such that not only the demand curve in the futures market shifts to the right but likewise does the futures supply curve. The number of contracts induced by demand for long positions has to grow continuously in order to cover increased oil supply. Hence, both demand and supply reactions to the high oil price are forces that tend to lower the oil price. A high price of oil while the level of stocks is high and effective demand is weak is a contradicting issue. This will sooner or later trigger counteracting movements of financial investors changing from net long to net short positions, which will lead the futures price and thus the spot price back to lower levels. Yet, it is impossible to generalize within which period of time and at which price level such a countermovement occurs, owing to complex reality and uncertainty, to wit, radical indeterminacy.

The strength of speculative impacts is likely to be affected by financialization. This issue has been intensely debated with respect to the market for crude oil. The degree of financialization is usually identified by the amount of open interest. It is unquestioned that open interest increased strongly since 2000 (CFTC, 2014). From time to time, financialization is said to be triggered by commodity market deregulation in the form of the US Commodity Futures Modernization Act (CFMA), which was signed into law in 2000. Kloner (2001) provides a short overview of this more juridical than economic debate. Some commentators consider the correlation between the oil price and open interest as a proof of how speculation in the futures market raises the oil price (see, for instance, Masters, 2008, pp. 5–6). Even though this conclusion may be premature and is often missing an adequate underlying theory, financialization deserves to be studied more in detail. It is often argued that total open interest has over time grown to a volume that is a multiple of global daily oil consumption. To infer from this finding that more oil is traded in the futures market than in the spot market with physical delivery is, however, wrong. Open interest is a stock variable and oil consumption is a flow variable. If open interest is coherently split according to the delivery dates of the futures contracts, open interest corresponding to a certain delivery day is only a fraction of daily oil consumption (Alquist & Gervais, 2011, pp. 3–4). Nevertheless, even though stock variables and flow variables may have a different nature, this does not allow for ignoring them. The larger the number of futures contracts of whatever duration, the more opportunities oil companies have to go short in the futures market rather than selling oil in the spot market. Larger open interest does not necessarily induce a higher price. Open interest may increase because oil firms prefer to raise their short positions. In this case, open interest is supply-driven and may lead to a lower price. Yet, as shown in Figure 2.1, the futures market is the channel through which speculation can exert an influence on the spot market. If demand increase is weak, the spot price increases only marginally. Thus, a sufficient amount of open interest is a precondition for a large demand increase for speculation to even be allowed to become effective. The comparison between open interest and daily consumption is not helpful. Speculation is expected to increase oil inventories, which reduces supply in the spot market. Hence, if the price elasticity of demand for crude oil is low, a relatively small increase in futures market demand suffices to raise the spot price even though open interest may only be a fraction of real oil consumption. Financialization in the form of growing open interest is thus not an automatic proof of a rising oil price but it enhances the potential for speculation to manipulate the spot market. We will argue that financialization becomes even more important when connecting it to the presence of expansionary monetary policy.

Based on the elaborated alternative view of financial markets, this chapter has applied these findings to the market for crude oil. As a crucial difference, there is the complication of the double nature of crude oil as a commodity and a financial asset that needs to be taken into account. But in accordance with stock markets, futures prices can change without changes in the real economy. Conversely, they can well exert an influence on fundamentals rather than merely be a reflection of the latter. This is in contradiction to neoclassical theory. While changes in the stock market are suggested to have an influence through Tobin's q and the wealth channel, the connection between the oil futures market and the spot market is much closer and more direct owing to the dual nature of oil. The alternative view takes complexity of real economic relationships into account and does not rely on a general equilibrium approach. It suggests dynamic interrelations between the real economy, the spot market for crude oil, financial markets or the futures market and monetary policy. They are analyzed in the next section in order to enlighten the effects that monetary policy has on the crude oil market.

2.2.2 Monetary Policy Effects Through the Real Economy and Financial Markets

The hitherto conducted analysis shows that monetary policy exerts impacts through economic fundamentals as well as through financial markets. The former is well known as the traditional way of how monetary policy affects the economy; the latter requires elaboration more in detail since it is due to the specific characteristic of crude oil as both a commodity and a financial asset. The mechanisms of the working of monetary policy will be emphasized later, when discussing its transmission channels. Fundamental and financial market effects both have distinguishing features that should be outlined first.

Effects on fundamentals
Neoclassical theory and our alternative view differ with regard to monetary policy insofar that mainstream economics assesses monetary policy impacts to merely concern the price level while hardly having any effects on real variables. Namely, it assumes a supply-side-determined equilibrium as given to which the economy reverts over the long run instead of accepting for both supply- and demand-side dynamics (Colander, 1996, pp. 28–29; Sawyer, 2002b). In the following, investigation of both supply and demand

sides will show that monetary policy may yield a much greater variety of outcomes than just a change in the price level. Monetary policy may be inflationary when it induces demand that exceeds production capacities. Otherwise, changes in the price level may be due to cost-pushes with regard to commodity price changes or altered wage levels. On the other hand, expansionary monetary policy may also lead to falling prices owing to lower interest rates entering production costs. Changes in the general price level leave the real price of oil unchanged and are therefore not the centre of interest in the current analysis of monetary policy.[16] In the cases of very high inflation rates or deflation, things may be different since they impede resource allocation that may give rise to additional effects on the price of crude oil (see Lavoie, 2006b, pp. 176–178). Since 1990, the OECD countries have featured moderate average inflation rates constantly below 8 per cent, and they have been falling over this time (OECD, 2014). For the moment, owing to these reasons, we abstract therefore from changes in the general price level as a result of monetary policy.

The endogeneity of money gives rise to the fact that the effectiveness of monetary policy depends on effective demand in the economy (Sawyer, 2002a, p. 42). The general way for a lower short-run interest rate to affect the economy is through an increased demand for credit. This can be credit for real investment, private and public consumption or, similarly, for mortgages (see, for instance, Gnos, 2003, pp. 325–326; Mishkin, 1995, p. 4). At a moment of monetary policy action, effective demand is given. New investment creates new demand since equipment has to be produced and real estate constructed. But likewise, new investment creates new supply by increasing production capacities (Lavoie, 2006b, pp. 178–190). Expansive monetary policy thus raises output by a small or large amount according to existing effective demand (see Forder, 2006, p. 232).

Total output is affected by monetary policy without there being a necessary impact on the general price level. This does not mean, however, that quantity effects caused by monetary policy are neutral to the structure of the economy. Some branches might react more strongly to a change in the interest rate level than other branches (see, for example, Keynes, 1930a/2011, p. 211). If the composition of total output changes, relative prices may change, too, even though the overall price level does not necessarily alter. In our examination, the relevant issue is how crude oil production or the oil industry reacts to monetary policy compared to the rest of the economy. The oil industry can be considered as the supply side of the oil market while the non-oil economy represents the demand side. If the non-oil economy grows more than the oil industry in response to a drop in the interest rate, oil demand rises more strongly than oil supply, which requires the oil price to increase in tendency. The term 'tendency' means that prices change to the extent that demand exceeds production capacities. If the utilization rate is below unity, stimulation by monetary policy raises oil production and output of the non-oil economy by a certain amount without the expectation of any price changes. In the case where the non-oil economy has a more robust response to monetary policy than the oil industry, the oil intensity of economic output shrinks. The opposite applies for the case when the oil industry reacts more to monetary policy actions than the non-oil economy. There are many indirect effects between the oil industry and the non-oil economy that take place simultaneously. Growth in oil demand induces an increase in oil supply and *vice versa* by conventional feedback mechanisms. The amount to which supply and demand-side reactions allow for the prevention of extreme price hikes depends on the magnitude of price elasticities of supply and demand.

Financial market effects
The above sections regarding financial markets in general and the futures market in particular lay the basis for the analysis of how monetary policy affects the economy or the market for crude oil through these financial mechanisms. Changes in the oil futures market take place through changes in the futures price and, analogously, the spot price. Thus, to make a preliminary summary, monetary policy exerts quantity effects through fundamentals and price effects through the futures market. Let us focus on the latter, that is, the price effect.

Monetary analysis exhibits that speculation is the mechanism through which monetary policy materializes in financial markets. But speculation requires a motivation of financial investors in order to effectively take place. They may have optimistic expectations that overshadow the effective demand constraint given in the real economy, but this begs the immediate question of where this optimism comes from. Hence, while it is not realistic to assume any speculative effects away by means of an efficient markets hypothesis, it is also unlikely that strong speculation activity comes out of nothing. Monetary policy is suggested to be a crucial factor.

Assume that the central bank cuts the short-run interest rate. Liquidity preference falls for the reasons already explained. A lower interest rate level changes investment perspectives. Bonds and bank deposits become less attractive. Investors would have to correct their profit expectations downwards. They increase demand for higher return and riskier financial assets in order to prevent a lower profit rate. Monetary policy acts as an exogenous force by setting the interest rate and affects liquidity preference in this way. Liquidity preference is based on the relative attractiveness of assets (see Bibow, 2006, p. 334; Lavoie, 2014, pp. 238–250). Interest rates influence that attractiveness, which reveals that liquidity preference is an endogenous issue. But from the point of view of the real economy and financial market performance, liquidity preference appears as partially exogenous, because it is not only an outcome of conventional market forces but also directly influenced by the central bank. Hence, liquidity preference can fall and speculation activity can accelerate even if there are no changes in the real economy and even if the economy is in a recession or even in a depression.

Profit purpose, wealth store and portfolio diversification are important issues that enable us to explain how a rising demand for financial assets spills over from the stock market to the futures market for commodities and specifically crude oil. Profit purpose is likely to be the main driving motive. Commodity indexes have been shown to face a lower standard deviation than conventional stock market indexes but with a comparable return over decades (Gorton & Rouwenhorst, 2006, p. 74).[17] This characteristic makes crude oil futures, either indexed with other commodities or individually, a welcome investment alternative. They are considered as a hedge against inflation (ibid., p. 75) and can thus serve as a store of wealth and for the diversification of the investment portfolio.

Basically, expansive monetary policy may lead financial investors to go short rather than long, which would bring them a certain return under given expectations, too. However, this scenario is not very plausible. First, lower interest rates used to have a positive effect on expectations about future economic performance, which tends to be linked, if at all, with a higher rather than a lower oil price. Second, it is probably easier for speculators to move the price upwards rather than downwards. Betting on a rising oil price without a change in fundamentals is easier done, because the simple reaction of oil companies is to raise short positions while a higher or lower quantity of additional inventories is accumulated. Speculating for a lower price would require oil producers to offer lower-price short positions even though they can sell oil at a higher price in the spot market. They certainly prefer the latter option. Third, inventories can accumulate infinitely but they cannot decline infinitely. At least when they are at zero, the price cannot be pressed down any further by raising oil supply beyond current oil production. Fourth, it is a fact that financial investors hold net long positions most of the time (CFTC, 2014). This empirical evidence is strengthened by the growing importance of index speculators (Masters, 2008, p. 6). Therefore, it is reasonable to associate lower interest rates in tendency with higher demand for oil futures long positions.

As we pointed out, each signing of a futures contract creates an unofficial form of money. It is backed by debt of an equal amount in analogy to conventional money creation. But since this debt does not bear any interest, no limits can be set to money creation in the form of futures contracts. The central bank uses the short-run interest rate as a tool to influence economic performance and to pursue its declared targets. Yet, this tool applies only to official money, which is created by commercial banks. From this point of view it may be argued that monetary policy does not have the power to affect the futures market, because unofficial money contained in futures contracts is fully independent of interest rate setting.

In fact, it is more appropriate to consider it the other way around. To participate in the futures market, an initial margin (complemented by the maintenance margin) is required to guarantee the liquidity of investors, because it is only official money that can serve as a means of final payment. The initial margin triggers the creation of an amount of unofficial money far larger than itself. Expansive monetary policy lowers financial investment cost. Investors have more liquidity available such that they can enhance their initial margins and thereby raise open interest. For clarity, assume for once that all futures contracts require immediate payment of the total contracted price instead of only the price difference that occurs over time. Financial investors would have to borrow much more interest-bearing capital. A given amount of capital could exert much less demand power in the futures market. Financialization of the commodity markets would be at a lower level. Yet, expansionary monetary policy not only boosts demand for futures by accelerating credit creation, it also triggers additionally a high leverage in the futures market, which raises hitherto existing futures demand by a multiple. Hence, the actual working of the futures market clearing system, and the nature of futures contracts, make monetary policy effects larger rather than smaller.

The alternative approach to financial markets and its application to the crude oil futures market can briefly be summarized by the following notion: speculation is the mechanism through which the futures market has an impact on the crude oil spot price. And it is monetary policy that drives this mechanism to become effective. Such a price effect does not require changes in fundamentals.

2.2.3   The Transmission Channels of Monetary Policy

For the analysis of monetary policy transmission channels in the context of crude oil, it is essential to be aware of the dual nature of oil. Hence, each channel has two aspects: a fundamental one and a financial one. The fundamental aspect comes into play when monetary policy affects crude oil as a real commodity. The financial aspect is relevant when the oil market is influenced by way of crude oil as a financial asset, that is, a futures contract.

The interest rate channel, the exchange rate channel, Tobin's q, the wealth effect and credit channels are the well-established channels generally identified in the theory of monetary policy. It would be wrong to take these channels as complete and perfectly complementary to explain the entire transmission mechanism of monetary policy. The theoretical identification of the individual channels took place at different times in the past (Mishkin, 1996, p. 1). Hence, they reflect different approaches and arguments that help detect the ways through which monetary policy is effective. This does not allow us to add up all channels to find the total impact of monetary policy. For example, the interest rate channel may contain features that are also taken into account by other channels. A part of monetary transmission runs the risk of being measured twice. The partial overlap of transmission channels requires a careful interpretation of the impact of monetary policy. Nevertheless, all channels are useful in the sense that they reveal many arguments and yield each important insight for the understanding of monetary policy transmission.

Monetary policy transmission through fundamentals

When monetary policy transmission mechanisms are discussed conventionally, it is their working through fundamentals that is meant. The debate is about how monetary policy influences both output and the price level through changes in investment and the consumption behaviour of firms and households. This examination is now made for the market for crude oil.

Interest rate channel The interest rate channel is often argued to be the principal one, since its influence on economic variables is quite direct and since it is the interest rate itself that is directly set by the central bank (see, for instance, Taylor, 1995, pp. 22–23). On the other hand, it is the most general and least specified channel and thus may as well be considered as a residual channel next to all other channels. As such, it is closely related to the other transmission channels and shares many basic characteristics with them. Some insights about the interest rate channel can therefore be applied to explain the remaining channels. Usually, the interest rate channel is described in the following way: an expansionary monetary policy leads to a lower level of interest rates, thus decreasing the capital cost, which raises investment. Higher investment means higher aggregate demand that results in higher output (Mishkin, 1996, p. 2). Yet, for our purpose it is not sufficient to know how the interest rate channel is linked to the whole economy taken as a single object. We are interested in the specific issue of the oil market and how it is influenced by monetary policy compared to the rest of the economy. Hence, our definition becomes broader: the interest rate channel depicts the way that the supply and demand sides in the economy are affected by monetary policy when all market participants react to an altered level of interest rates.

Interest cost is a part of total production cost. Falling interest rates allow firms to finance production at better conditions and thus to produce at lower cost, which transmits to lower prices, generally speaking (Sawyer, 2002a, p.42). Lower prices will raise demand. The firms, in expectation of higher profits to realize, raise the quantity produced. To the extent that effective demand allows for higher returns, firms take additional credit at the more favourable interest rate and enlarge their supply capacities by productive investment. This is the supply-side reaction of the economy (see Lavoie, 2006b; Sawyer, 2002b).

On the demand side, it is the investment of firms that exerts additional demand by purchasing new equipment. For private households, consumption wishes can be satisfied by getting indebted more easily when the central bank lowers the interest rate. Similarly, for savers the opportunity cost of consumption decreases because savings account returns fall, too. For instance, serving interest payments of a mortgage loan then take a smaller share in future income of a household. Total demand increases therefore when monetary policy is expansionary. If rising demand has a significant effect on the price level at all, it leads to rising prices. It is argued that real estate and consumer durables represent principally investment-like expenditures for production equipment (see, for instance, Mishkin, 1996, p.2). This is true in the sense that all expenditures create effective demand in the same way. However, it is essential to distinguish supply and demand sides and their responses to monetary policy. All expenditures on the market for produced goods and services contribute to effective demand and thus raise output. But not all expenditures raise production capacities. Hence, the character of expenditures has an influence on total production capacity, on the degree of capacity utilization and thus on the price level. This shows again that, given a certain monetary policy action, it is not possible to predict the change of prices. Therefore, money cannot be the object that is itself responsible for inflation.

Whether the interest rate channel has a stronger or weaker effect on the oil industry than on the non-oil economy depends on various factors on both the supply and demand sides. All product prices, including the oil price, consist of a number of components. A price divides into production cost and a profit share. Production cost is composed of equipment depreciation (including machines, real estate, and so on), expenditures for commodities and other raw material, salaries and interest payments.

Interest cost is unavoidable for a company, because it requires capital to finance production. Depending on the performance of a specific company, borrowed capital takes a larger or smaller share in total capital. Hence, a decrease in the interest rate lowers production cost to the extent that the production of a company is financed by borrowed capital instead of equity. A higher share of borrowed capital makes the interest rate channel more effective.

The profit share is a two-sided issue. The higher the profits, the more companies are able to finance production out of their own resources and thus the share of equity is higher. On the other hand, assuming that the profit share is a mark-up of a certain percentage rate of production cost, as is often done in many models of various origins (see, for example, Blanchard & Illing, 2006, p. 194; Kalecki, 1987, p. 104), the larger the production cost, the larger the profit. A lower interest rate then not only lowers production cost but also absolute profit per production unit. The total effect of monetary policy on the price level is thus enhanced. The degree of competition that influences the size of the mark-up is therefore a further influencing factor. The crude oil market is a distinguished case in this respect. OPEC produces about 40 per cent of total crude oil (EIA, 2015b). Even though its effectiveness in agreeing on production quantities is often doubted (see Smith, 2009, pp. 151–154), its existence is to be taken into account as it probably raises the profit share in the oil price and the possibility of self-financing investment.

These supply-side considerations concern both the oil industry and the non-oil economy. In general, investment is an expenditure and as such represents demand. However, what is crucial in the longer run is that it enhances production capacities and therefore *ceteris paribus* increases supply. Perspectives change when oil production is considered. Expenditures outside of the oil industry (as an average, ignoring now the variations in the oil intensity of expenditures) raise demand for oil even if they are an investment of the type described above and hence would appear as an increase in supply from the point of view of the whole economy.

For the reasons emphasized here, the interest rate channel tends to give rise to higher rather than lower absolute oil production in the course of expansionary monetary policy, since both supply and demand sides are positively affected. The quantification of the effect in comparison to the rest of the economy is less clear. It depends on the relative strength of monetary policy impacts on the oil industry and on the non-oil economy. For an analytical proceeding, we start again from the equivalence of supply and demand in the oil market. Oil supply and oil demand are equated by the oil price in every moment of time. The interest rate channel does not distort this equivalence, but it alters some relevant variables. When demand (left-hand side) changes, either supply or the oil price (right-hand side) or both have to change, too. This is illustrated by equation 2.2:

$$\Delta(consumption) + \Delta(non\text{-}oil\ investment) = \Delta P_o + \Delta(oil$$
$$industry\ investment) + \Delta(capacity\ utilization) \qquad (2.2)$$

Equation 2.2 is only a rough approximation of the supply–demand equivalence in percentage changes and covers the variables that are relevant with regard to the interest rate channel. Consumption expenditures, that is, purchase of real estate and consumer durables and non-durables, and non-oil economy investment, represent the demand side in the oil market. If one or both demand components increase, the supply side responds either by investing in equipment or by raising the capacity utilization rate. If the oil industry does not react at all or only insufficiently, the oil price has to increase in order to re-equate supply and demand in the oil market. The reason why the price variable as the balancing variable is on the right-hand side instead of the left-hand side of equation 2.2 is that both supply-side components are technical variables. They determine actual physical supply measured by quantity rather than price. In contrast, the demand components express purchased quantities measured in units of money. Hence, they are monetary variables. They divide into a price variable and two supply variables.

Equation 2.2 reveals two effects of the interest rate channel, that is, changes in production cost and in investment behaviour. If a drop in the interest rate level lowers production cost and hence prices more in the oil industry than in the rest of the economy, ΔP_O is negative. This triggers an increase in oil demand. The left-hand side of equation 2.2 is therefore at least weakly positive. Oil supply responds by either increasing additional investment or by an augmentation in capacity utilization. Oil supply grows more than oil demand owing to the drop in P_O. This is due to the logic of the equation as well as to increased profit expectations of oil companies outlined above. The effect might be dampened by a partial re-increase of the oil price due to higher demand for oil.

In the same way, if investment in the oil industry triggered by a lower interest rate is larger than the sum of investment in the non-oil economy and consumption growth, oil supply *ceteris paribus* grows larger than oil demand. This requires the oil price to fall. Alternatively, capacities increased by investment may be evened up by lower capacity utilization. However, this is not to be expected, since it is unlikely that companies invest without using created capacities.[18] In both these cases – a fall in relative oil production cost and a relative rise in oil industry investment – oil production increases more than output in the rest of the economy. The lower oil price re-equates both sides. Higher oil supply therefore transmits to higher oil consumption. Oil intensity of total output is now larger than before.

Conversely, if oil production cost falls by less than production cost in the non-oil economy, the oil price increases and demand falls relatively. Equally, if oil industry investment grows less than investment in the rest of the economy, oil supply tends to rise to a lesser degree than oil demand. Given the case that oil industry investment is fixed, increasing demand requires a higher rate of capacity utilization. If capacities are fully employed, the oil price starts growing. This means that increased oil demand translates to a higher price while supply is constrained. The oil intensity of the economy shrinks in this case.

In general, we have to distinguish between two different types of inflation, that is, demand-pull and cost-push inflation (see, for instance, Rochon, 2004, pp. 8, 19; Setterfield, 2006; Vernengo, 2006, p. 471). The term 'inflation' might go too far, because our interest is on the oil price rather than the general price level. Yet, it nevertheless serves to describe this particular case. Demand-led inflation occurs when demand exceeds production capacities. Cost-push inflation takes place when production costs rise and transmit to prices. Gibson's paradox is an approach that explains price changes from the cost perspective. Since interest is a part of production cost, expansive monetary policy lowers prices through a reduction of production cost. Demand-pull inflation is the preferred explanation of neoclassical theory, where cost aspects play merely a marginal role (Barro & Grossman, 1974; Brunner et al., 1973; Gnos & Rochon, 2007, p. 374). In our investigation, both approaches should be taken into account. Cost-push inflation has been examined by the decomposition of price components. As regards the first principal effect, a cut in the level of interest rates lowers production cost and therefore represents a negative cost-push. In the case where oil industry investment reacts more to a cut in the interest rate than the rest of the economy, the oil price is expected to fall owing to rising supply. In principle, one may consider this reaction as a negative demand-pull. However, this phenomenon is more likely to be linked to another (negative) cost-push. It shows that the two aspects of the interest rate channel, that is, lower production cost and higher investment, are tied together in many cases. Investment does not grow out of nothing. Oil companies decide to invest if profit expectations are bright. Assuming away strategic behaviour, they raise investment if they know that it will augment profits. Hence, newly created capacities are used and effectively raise oil production. The profitability condition is sufficient effective demand. The increased quantity of oil must therefore be supplied at a price low enough for existing effective demand to clear the market. Companies only accept a lower price if production cost is lower, too. For a given level of effective demand, the interest rate channel raises investment in the oil industry only if the lowered interest rate reduces production cost sufficiently. Increased oil supply that is induced by an expansionary monetary policy is therefore, if not identical, at least closely linked to a negative cost.

In the opposite case, where the oil industry reacts less to an expansionary monetary policy, the cost-push is directly effective, too. But since production cost falls less than in the non-oil economy, the cost-push is positive in relation to production cost in the rest of the economy.

The second principal aspect, investment behaviour, refers to a classical type of demand-pull inflation. As mentioned, the oil price only increases if excess demand rises above production capacities. Hence, while the price is expected to fall if oil industry investment grows relatively stronger, the opposite price effect is more ambiguous if non-oil investment and consumption demand grow more. One should be aware that what appears as a demand-pull in the oil market may be the counterpart of a cost-push in another market. Strong investment in the non-oil economy may be due to lower production cost. Thus, a negative cost-push in the non-oil economy indirectly creates a demand-pull in the oil industry.

The issue of monetary policy, quantity effects and price effects is complex. The separation of changes in production cost and investment behaviour does not suggest that they are easy to separate in reality. They are not only tied together empirically but share many similarities in theory as well. In the greatest part of economic literature about monetary policy transmission, the economy is investigated as a whole. The inspection of a single economic branch additionally complicates the issue by introducing relative rather than only absolute variable changes within the whole economy. Relative quantity and price effects reveal how mutual impacts within the economy work. Individual effects tend to be overlapped by other cost-pushes or demand-pulls. Expectations under uncertainty may induce investments that yield different final results.

Exchange rate channel Monetary policy transmission through the exchange rate is a well-recognized channel and has a specific connection to the price of crude oil. A cut in the interest rate by the US central bank lowers the return of holding deposits in US dollars relative to returns of other currencies. Demand for dollars decreases and so does its exchange rate against other currencies.[19] Depreciation makes US exports cheaper abroad and raises the price of US imports. Net exports increase and so does US output (Mishkin, 1996, p. 5). The US dollar and the price of crude oil share a specific feature assessed in a cointegrating relationship.

US dollar depreciation, against the euro for instance, leads to a higher oil price while an appreciation simultaneously lowers it (Zhang et al., 2008, pp. 981–982). World trade in crude oil is denominated in US dollars. A lower value of the US dollar threatens profits of oil companies producing outside of the United States. They raise the oil price to avoid shrinking revenues measured in their currencies. In the same way, oil demand from outside of the United States increases, because oil can be purchased more cheaply owing to the weak US dollar.

Determining the effects of a change in the US dollar exchange rate against, say, the rest of the world from a global point of view is more complex. In addition to the already mentioned reactions of foreign oil production and oil demand, a depreciating US dollar lowers domestic oil demand and raises domestic oil supply on the one hand. On the other hand, boosting exports of goods and services increase oil demand while decreasing imports reduce oil demand.

The definitions of domestic and foreign country change when some countries decide to peg their currency to the US dollar. For instance, this applies for six oil-producing countries in the Arabian Gulf (Cevik & Teksoz, 2012, p. 3). Such shifts in borders of pegged currency areas alter supply and demand effects. These partial effects have supplementary and adverse impacts on the price, production and consumption of crude oil. The direction of the resulting net effect can hardly be identified and depends on a considerable number of price elasticities of supply and demand. On the other hand, Lizardo and Mollick (2010, p. 405) observe that a higher oil price depreciates currencies of significant oil importers like Japan relative to the US dollar. This means that a US dollar depreciation against an average world currency basket is partially reversed and mainly takes place relative to significant oil exporters and large economies where oil imports have a lower share in total imports, such as in the euro area. What remains from the depreciation in the US dollar exchange rate, in spite of all additional impacts, is the higher oil price. These findings suggest that the exchange rate effect on the oil price is well established. In contrast, effects on crude oil quantities, that is, production and consumption, are more ambiguous.

Tobin's q   Even though we are still analyzing the fundamental aspects of transmission channels, the role of financial markets in monetary policy transmission is already involved by the approach of Tobin's q. Tobin (1969) developed a coefficient, q, which aims at capturing the effect that a change in the level of interest rates has on investment behaviour by ways of companies' capital prices. The coefficient 'q' is defined by the ratio of the market value of a firm's capital to the replacement cost of that capital.

The market value is given by the liabilities that the firm commits to in order to finance production multiplied by the market price of liabilities. Replacement cost of capital has to be estimated by the cost that would have to be afforded if the firm had to substitute new production equipment for the existing equipment. The higher the ratio, the more attractive is the condition at which a firm can raise new capital. If q is larger than one, a company benefits from issuing new stocks or bonds, because it can obtain capital at lower cost than its market value will be. Thus, the firm is induced to increase investment. Variation in q can have numerous causes, one of which is monetary policy. A cut in the interest rate leads investors and savers away from deposits to alternative investment opportunities. Demand for financial assets rises and so does their price. Hence, an expansionary monetary policy raises investment by an increase of Tobin's q.

A varying value of Tobin's q is incompatible with the efficient markets hypothesis. In a world where prices of financial assets move exactly along fundamental developments, q should be one (Tobin & Brainard, 1990, pp. 547–549). Anything else would imply that factor allocation is not efficient. As mentioned, the existence of Tobin's q as a transmission channel of monetary policy requires at least nominal rigidities, imperfect information or other market imperfections. Even though Tobin (1969) justifies a flexible q by means of a general equilibrium model and exogenous money, he applies a central feature that is also included in the concept of endogenous money and in our alternative view: the interest rate is set exogenously (Tobin, 1969, p. 26). Demand and supplies of assets therefore adapt to monetary policy. Tobin (1969) does not elaborate further the relationship between assets and real output but at least assigns to the financial sphere a certain independence from fundamentals (ibid., p. 16). This is a necessary condition for the financial market, and hence Tobin's q, to have any influence on real variables.

The higher the price of stocks and debt of a corporation, the higher this corporation is valued by the market. The prices of these financial assets must be variable so that the value of q is allowed to change over time. A condition for price variation is tradability of assets. A firm credit that is not securitized and therefore not tradable in the stock exchange will always have the same price, which is equal to the nominal value of the debt. Tobin's q of a small firm that does not issue any securities and whose production is financed only by bilateral credit from a bank has a constant value of one. A larger share of tradable securities thus implies a larger variation of q over time. The more a company is financed by tradable stocks and bonds, the more monetary policy can become effective through the transmission channel of Tobin's q.

The market for crude oil exhibits large corporations, almost all of which are listed on the stock exchange. Even though not empirically assessed yet, this opens the potential for Tobin's q to have an impact on the oil market. In comparison to the rest of the economy, this channel is likely to be more effective in the oil industry. We know without investigating the structure of the non-oil economy in detail that a large share of its output is produced by small firms not quoted on the stock exchange. This is in contrast to the oil industry, which consists for the most part of large companies.[20] From a macroeconomic perspective, the share of tradable securities should therefore be larger in the capital of the oil industry than in that of the non-oil economy. Hence, q grows stronger in the oil industry than it does in the rest of the economy in response to an expansionary monetary policy. This induces relatively stronger investment in the oil industry. The consequences should then be those of the corresponding case analyzed with regard to the interest rate channel: higher investment in the oil industry than in the non-oil economy is likely to lower the oil price and to raise the oil intensity of the economy. Like a lower interest rate, easier capital acquisition owing to a higher q is reflected in a lower production cost.

Wealth effect   In some sense, the wealth effect is for private households what Tobin's q is for firms. While the latter raise investment in response to an increase in q, the former augment consumption when the market value of their assets increases. This transmission channel is based on the argument by Ando and Modigliani (1963) that consumption not only depends on current income but also on current assets and expected future income, that is, total lifetime wealth. The asset portfolios of households may contain stocks and bonds but are usually dominated by real estate (Rossi, 2008, p. 265). By the same mechanism as with Tobin's q and as examined in the alternative approach to financial markets, an ease in the conduct of monetary policy raises demand for assets. Higher prices of financial assets transmit to real assets directly or indirectly. The case of real estate is similar to crude oil, since it is not only a real asset but also a financial asset in many cases. A higher price of assets makes households feel richer and thus they raise consumption. The strength of this particular effect is quite controversial (see Ludvigson et al., 2002).

From equation 2.2, which shows how monetary policy affects the oil market from supply and demand sides, it becomes obvious that the wealth effect is exclusively a demand factor. In the course of a drop in the interest rate level, households raise their expenditures for consumer durables or real estate and thereby indirectly increase demand for oil. The wealth effect is not supply-driven, because it does not contribute to the increase of production capacities. Hence, the wealth effect leads to higher change of demand for oil relative to oil supply and thus tends to raise the oil price depending on spare capacities in the oil industry. In the same tendency, oil intensity of economic output falls.

Credit channels   The credit view is an extension, specification as well as critique of the traditional interest rate channel. It considers that it is not merely the level of interest rates as such that transmits to the economy and affects consumption and investment decisions. It is rather more that monetary policy in the same way also alters the conditions for firms to have access to capital. In this sense, it is an 'enhancement mechanism' of the conventional transmission channels (Bernanke & Gertler, 1995, p. 28). The literature distinguishes between the bank lending channel and the balance sheet channel, which are also called the narrow and broad credit channels.

The central requirement for the existence of credit channels is that the external and internal financing of firms are not perfectly substitutable (Rossi, 2008, p. 266). External financing means borrowing in the credit market while internal financing occurs in the form of firms' own profit reinvestment or emission of new stocks and corporate bonds. According to the assumption, external and internal financing are subject to different conditions. Some companies do not have access to capital markets and thus exclusively rely on bank credit. The lender is not fully informed about the financial situation and creditworthiness of the borrower (see Stiglitz & Weiss, 1981). Owing to this asymmetric information, the loss risk for a bank by granting a loan is larger. Hence, it imposes a premium in addition to the given general interest rate in order to compensate for the risk. The premium is variable. The higher the general interest rate level, the higher is the risk of an investment failure and higher, too, is the loss risk of the bank.

The bank lending channel takes the perspective of the lender, that is, the bank (see Kashyap & Stein, 2000; Kishan & Opiela, 2000; Stein, 1998). Owing to the basic assumption, the firm does not have the unrestricted possibility to change from bank credit financing to self-financing or emission of financial assets in response to a change in the interest rate. This allows for emphasizing the entity of the commercial bank as a loan supplier that is different from other financing sources. Thus, the firm's capital volume is constrained by the bank's credit. Even if money is demand-determined and hence basically not limited by the supply side, the issue of loans is constrained by the creditworthiness of the borrower. A further condition for the bank lending channel is, first, the requirement for banks to hold minimum central bank reserves and, second, the impossibility to avoid these requirements by liability management (Rossi, 2008, p. 270). An expansive direction of monetary policy is effective either by lowering the share of required reserves with respect to the total volume of loans or by making access to reserves easier, or both. A loosened constraint allows the bank to raise the issuance of loans. Lower interest cost decreases the risk of investment projects such that the risk premium shrinks, too. The bank lending channel thus strengthens the impact of the interest rate channel.

Unlike other countries, banks in the United States are required to hold positive minimum reserves. These reserves make banks incapable to act unless they hold sufficient liquidity.[21] Thus, the first condition is fulfilled. Indeed, banks that hold little liquidity are significantly more affected by changes in monetary policy, because the loosening of the reserve constraint has a greater impact on them (Kashyap & Stein, 2000, pp. 408, 425). As a limiting influence, however, increasing financialization in the last decades is likely to have facilitated liability management. It allows enhanced money and credit creation by banks without violating their reserve requirements. The reaction of bank lending to a change in monetary policy is thereby reduced. Yet – and this may also be important as an influencing factor on the price of crude oil – financialization and the bank lending channel seem to have a double-edged relationship. Circumvention of reserve requirements leads to larger credit volumes and risk exposure. Shadow banks do not create money by loan issuance but rely on financial resources that have their origin in the money creation process of conventional banks. The interconnectedness of shadow banks and traditional banks can amplify systemic risk in the course of growing financialization (Financial Stability Board (FSB), 2013, p. 21). A higher leverage deepens bank troubles in times of crisis and thereby financial instability, and thus worsens banks' liquidity situation. The cost of deposit funding is more stable owing to the stability of the deposit rate of interest than the cost of volatile funding in the market by the emission of securitized financial assets. Banks that fund themselves to a high degree by market sources reduce credit supply more strongly in crisis periods than other banks (Gambacorta & Marques-Ibanez, 2011, pp. 15, 17). The dependence on monetary policy action becomes larger. Hence, the bank lending channel plays an increased role in loosening liquidity during periods of financial instability. Ironically, this appreciation may be promoted by financialization, which is suggested to have contrary influences in normal times.

The balance sheet channel approaches credit effects from the point of view of the borrower. It not only concentrates on commercial banks as a source of credit but also includes market funding. The effect of a change in interest rates is augmented by two variables: firm cash flow and collateral (see, for instance, Angelopoulou & Gibson, 2009; Bernanke & Gertler, 1995; Bernanke et al., 1996). A cut in the interest rate lowers firms' financing cost, especially if it is short-run financing. In response, cash flow increases by a considerable amount in the short run. This improves the financial situation of a firm and mitigates the loss risk of the lending entity. The finance risk premium, in addition to the general interest rate, decreases. Firms react by increasing investment expenditures. Better financial conditions of firms raise their market value and thereby the net worth of assets that serve as collateral in credit agreements (Bernanke & Gertler, 1995, pp. 35–36, 38–39). Firms can offer more and higher-worth collateral, which reduces the lender's risk and has a decreasing effect on interest cost. Again, the effect of monetary policy is assumed to be weaker, the easier companies have access to internal financing. The approach therefore takes the heterogeneity of borrowers into account. The balance sheet channel is usually found to be significant and more important than the bank lending channel (see, for example, Aysun & Hepp, 2013). This is not a surprising result since the bank lending channel is argued *a priori* by theory to be narrower and to cover fewer features than the balance sheet channel. As with the bank lending channel, financialization is likely to make the balance sheet channel more important when a financial crisis occurs.

The credit view from the borrower side usually concentrates on firm borrowing. Even though this is not done regularly, balance sheets of private households should consequently also be taken into account. An expansionary monetary policy raises financial asset prices. This improves the financial and liquidity situation of households and reduces their risk of getting into financial distress (Mishkin, 2001, p. 4). Households can afford more consumption and are imposed a lower premium for external financing by banks. Credit volume for consumption expenditures increases. The mechanism is closely related to the wealth effect.

The structure of the oil industry, mainly consisting of large corporations, implies that it has good access to internal funding sources. This is in contrast to small firms, which are suggested to fully depend on bank credit. By the same argument that states a stock exchange listing of oil companies as above average, the oil industry has a better access to capital markets than the rest of the economy. Thus, under normal circumstances, we expect investment in oil production to react less to a change in monetary policy than expenditures in the non-oil economy. According to equation (2.2), this implicates a stronger reaction of oil demand relative to oil supply and requires a higher oil price in tendency. However, the capital structure of the oil industry may be riskier than that of the non-oil economy. The balance sheet of oil companies directly depends on sales prices of their sales product, to wit, crude oil. It strengthens when the oil price climbs high and contracts when it falls. Oil companies' capital funding may be less constrained by monetary policy in boom periods, when the oil price tends to be high. Conversely, the constraint becomes tighter when financial instability occurs, owing to a falling oil price. The tighter the borrowing constraint, the larger is the effect of a change in the interest rate by the central bank. Financialization is suggested to exacerbate fluctuations not only of the oil price but also of the financial performance of the oil company. Effects of monetary policy on the price of oil thus affect the balance sheet channel by an additional mechanism. This issue is revisited in the next chapter as part of the investigation of the price effects of monetary policy in connection with financial markets.

In a given situation where capital availability for the oil industry and the balance sheets of oil companies are less constrained than those of the non-oil economy, the latter expands investment relatively more, because it is more strongly affected by monetary policy. This means that oil demand increases in strength compared to oil supply. This effect tends to raise the oil price and to lower the oil intensity of the economy. In another situation, for instance in a financial crisis, the oil industry's access to credit is likely to be more constrained. The opposite effects should occur then.

Monetary policy transmission through financial markets
The fundamental aspects of monetary policy transmission are about the real economy, where oil serves as a physical consumption good. The chapter at hand is concerned with the monetary policy transmission to crude oil when it is traded as a financial asset instead of a commodity. It is shown that this aspect leads to mere price effects in contrast to quantity effects through fundamentals.

A change in the interest rate by the monetary authority has its effects on supply of and demand for financial assets as argued above. A lower interest level decreases liquidity preference and thus increases demand for financial assets even if the economy is stagnating. Crude oil futures contracts represent an asset, too. Increasing demand for futures implies purchases of long positions. Their price is likely to rise in the course of an expansionary monetary policy, which transmits to the spot price of oil. In opposition to the mere consideration of fundamentals in the previous section, the financial market effect through the interest rate channel implies that it is not only firms or consumers who react to an altered interest rate but also financial investors. Likewise, there is not only investment in real assets in the form of equipment, real estate, and the like; investment can also take place in the form of purchases of financial assets in general and of crude oil futures in particular.[22] The financial aspect of the interest rate channel shows that financial investment and hence the oil price increase. The fundamental part of the channel leads to a certain quantity effect by influencing oil production and consumption, while the financial market aspect exclusively yields a price effect.

In the same way that the exchange rate channel is different from the point of view of fundamentals, it differs from the other channels when considering its financial market impacts. Assuming that financial investors know that the oil price and the US dollar exchange rate are correlated, they build their corresponding expectations. When the US Federal Reserve System (Fed) lowers the interest rate, which leads to a depreciation of the US dollar, investors anticipate a higher oil price and start betting by purchasing oil futures long positions (Zhang et al., 2008, p. 982). Consequently, the oil price increases.

Similarly, when an expansionary monetary policy raises Tobin's q of firms, the improved opportunity to invest does not necessarily lead to the purchase of production equipment. Emission of new stocks and corporate bonds under favourable conditions raise the capital needed for either real or financial investment. Moreover, a higher net worth of companies induces riskier behaviour concerning financial investment. If a certain share of additional financial investment goes to the crude oil futures market, the oil price increases. Financial market transmission through the wealth effect is analogous.

The credit channels have the same price effect. An expansionary monetary policy loosens the constraint of banks' and companies' balance sheets. Increasing issuance of bank loans serves to fund financial investment. Improving balance sheets of companies allow them to get more credit granted. They decide whether to invest it in real production equipment or in stocks, bonds and futures contracts. One should also consider the specific relationship of oil companies to the balance sheet channel. The effects of transmission channels through financial markets alter the price of crude oil and, in doing so, the latter affects the balance sheets of oil producers, whose oil inventories and sales represent an important part of real assets. Their cash flows and collaterals improve, therefore, not only by means of a lower interest rate but also by the financial market effect owing to the dual nature of oil. An expansionary monetary policy is therefore likely to relax the financial situation of oil companies more than that of the rest of the economy. In contrast, a contractionary monetary policy constrains oil companies' balance sheets more and threatens their creditworthiness more than those of the non-oil economy.

2.2.4 The Interaction of Fundamentals and Financial Markets

Obviously, the fundamental and financial market aspects of monetary policy transmission are related. The quantity effects of monetary policy through fundamentals are ambiguous since it is not completely clear if oil quantities increase relative to the rest of the economy and, hence, if there is a change in the oil intensity of total economic output. The interest rate channel does not give clear *a priori* results whether the oil industry or the non-oil economy reacts more. The exchange rate channel is an exception in the sense that it represents a mere price effect rather than a quantity effect from a global perspective, whereas the price effect is unambiguous. Tobin's *q* of the oil industry is likely to respond relatively stronger to monetary policy than that of the rest of the economy. In contrast, the wealth effect is merely a factor that enters the economy from the side of the non-oil economy, since it raises consumption and thus demand for oil. The credit channels allow even less for a clear statement about relative strengths in responses to monetary policy. They depend on the state of the business cycle. Whether the oil industry or the non-oil economy, that is, the oil supply or demand side, reacts more to monetary policy is determined by the complex aggregation of the channels of transmission through fundamentals. What can be said is that monetary policy can affect fundamentals of the economy and in particular of the market for crude oil without a reason to expect a significant price change of a certain sign. Given a certain change in oil production and consumption following a change in monetary policy, existing spare capacities and capacity-increasing investment tend to prevent a climb in the oil price.

On the contrary, price effects resulting from monetary policy transmission through financial markets are unambiguous. Each particular transmission channel contributes to a rising oil price when monetary policy is expansionary. This occurs since financial investment tends to enter the futures market on the demand side by purchasing long positions. Once the direct effects of monetary policy on fundamentals and futures market are assessed, many interactions emerge that have to be taken into account. We highlight them against the background of a cut in the interest rate by the central bank. Let us start with the influence that transmission effects through fundamentals have on the futures market. Quantity changes in oil market fundamentals that take the form of higher demand by consumption expenditures, higher non-oil production or higher oil industry investment can basically result in almost any possible oil price level. We concluded above that this case of indeterminacy does not raise expectations about significant systematic price changes. Given that there is no price change, financial investors do not adapt their asset portfolio, which they have chosen in response to a change in monetary policy. In the potential case of a drop in the oil price, either due to large investment or a strong negative cost-push in the form of lower interest rates, speculators hesitate to bet on a price increase and thereby purchase less long positions. In the opposite case of a price increase, motivation for financial investment grows even stronger and net long positions are likely to increase. While the financial market price effect of monetary policy is weakened in the former case, it is strengthened in the latter.

Next, let us have a look at the effect of monetary policy transmission through financial markets on oil market fundamentals. The one-directional effect of financial markets on the oil price has its impacts on oil market fundamentals that are more pronounced since its sign is quite unambiguous. A higher oil price caused by financial investment lowers demand and increases supply in the spot market. Futures market trade represents an exogenous influence factor to supply and demand forces in the spot market. Figure 2.1 showed how financial speculation affects the oil spot market without preceding changes in fundamentals. Consequently, a thereby accelerated price does not necessarily revert back to its initial level when supply and demand in the spot market adapt to this exogenous impact. Moreover, we explained that a higher oil price induced by financial markets can also be profitable for oil companies even though it reduces demand in the spot market. Decreasing oil sales are overcompensated by increasing open interest in the futures market, which is, measured in US dollars, equivalent to spot sales from the point of view of oil companies. Hence, the oil price is not inevitably reduced even if oil demand decreases. As a usual reaction to higher prices, the oil industry increases supply by extending production capacities. This will pull the oil price down over a longer period despite its current high level. How much oil supply is increased by the oil industry, how low the oil price eventually falls and within which period of time depends on the relevant elasticities as well as on the strength of the financial market impact and its continuance. Persisting expansive monetary policy actions have the potential to keep the oil price high for a longer time period despite counteracting movements in fundamentals.

The crucial difference between price climbs induced by speculation in the futures market and those caused by excess spot demand can be explained as follows. Assuming away financial markets, the oil price increases either when supply decreases while demand is stable or when demand increases while supply is stable, or both. In any case, supply fails to satisfy existing demand at the existing price. In contrast, a speculation-caused price increase is driven by demand from the futures market. It is not that spot demand grows in excess of spot supply. Neither is it that a supply constraint in the spot market is the cause of the higher price, since oil production is confronted with spot demand that is falling rather than rising. As a consequence, oil companies respond by higher investment and raise production capacities. But increased oil production does not necessarily lower the oil price immediately as would be the case if excess demand in the spot market was the origin of the price hike. Owing to the exogenous demand force, that is, demand for futures contracts, the oil price may last at the higher level even when demand in the spot market decreases. Once monetary policy becomes less expansionary, the price effect of financial markets starts disappearing and the oil price necessarily decreases.

The fact that the financial market effect involves higher oil production through oil industry investment is rather intuitive. The influence on oil consumption is less obvious and should be emphasized. Even though oil supply and oil production on the one hand, and oil demand and oil consumption on the other, are often taken as practically the same thing, they differ in some important aspects. Supply and demand extend to the futures market, while production and consumption are limited to the spot market. When expansive monetary policy raises the oil price by means of speculation, both oil supply and oil production increase because they react to the same incentive. Total oil demand increases, too. However, it increases not uniformly as its composition changes. Demand in the spot market shrinks but is replaced or overcompensated by demand in the futures market. Falling spot demand means falling oil consumption. Growing total demand and decreasing consumption are reflected in rising inventories as has been explained in detail in this chapter. Speculation contributes to falling oil consumption whereby the extent of the reduction depends on the price elasticity of (spot) oil demand.

At the present point of analysis, neoclassical theory would probably argue that the financial market effect, if it ever exists, has its distorting influence on fundamentals before it vanishes again. The economy goes back to its long-run equilibrium path. However, there are many reasons to suggest that monetary policy transmission through financial markets has long-term persisting impacts. On the supply side, one might suggest that rising oil production will go back as soon as the oil price falls back owing to the fading out of the financial market effect. A lowering oil price will reduce oil supply by the same amount as it had raised supply when it was increasing. Yet, to be realistic, real investment in oil production is hard to be withdrawn once it is realized. Since a large share of it consists of fixed cost expenditures, oil sales are required to cover fixed cost in addition to the variable production cost. Constructing a whole new production plant might be unprofitable when the oil price is low. In contrast, given that the production plant exists already (considering fixed investment as a sunk cost), producing oil is profitable as long as the return of a unit exceeds its variable production cost (see, for instance, Vickers, 1960, pp. 407, 409). Thus, existing increased production capacities tend to be used even if the existing lower level of the oil price in a given moment does not impel any investor to raise oil industry investment owing to missing profitability. Therefore and crucially, we conclude at this point that real oil industry investment induced by the high price, which is itself due to monetary policy and speculation, lowers the oil price again and probably to a lower level than at the beginning of the causal chain due to larger production capacities. At this point, the economy ends up with a higher oil production than before.

Oil demand may also be argued to go back to its initial level after the disappearance of the financial market effect. But there is one aspect that has so far been in the background, and one that mitigates the hitherto conclusion. We have assumed that technology and technological progress are given in a specific point in time, so that the effects of monetary policy can be examined without having to take into account such long-run developments. Here now, the neutrality of technological process should be abandoned. It may be true that consumption expenditures react symmetrically to oil price changes in the sense that oil consumption is the same before a price climb as when it drops back to its initial level. For example, consumers might on average buy smaller and more efficient cars when the oil price is high and move back to more wasteful motors after the reversion of the price.[23] In contrast, it is more likely that in many areas, such as industrial production or house heating, producers and owners start looking more intensively for alternatives. This should promote technological innovations in favour of more efficiency and non-fossil energy sources. For instance, Bayer et al. (2013) show that a high oil price has a significant positive influence on renewable energy innovation. Once new production technologies are invented, applied and established, they are likely to sustain even after the fall of the oil price. The demand curve then shifts downwards; oil consumption is *ceteris paribus* lower than before the price climb. This effect is less obvious and more unforeseeable than supply-side effects. So it should be considered as an aspect that mitigates the hitherto arguments but does not replace their basic logic. Since the share of renewable energy is below 10 per cent of total global energy consumption, these demand-side effects are empirically limited (EIA, 2015b). If real investment in the oil industry, triggered by the speculative increase of the oil price, lowers that same price to a lower than initial level, oil consumption is likely to be higher than initially as well. The oil intensity of the economy therefore increases. This issue will be subject to central discussions in the empirical and economic policy sections of this work in Chapters 4 and 5 respectively.

The price effects that emanate from the oil futures market alter supply and demand in the spot market and can therefore have lasting impacts on fundamentals. The time difference of monetary policy transmission through financial markets and through fundamentals additionally supports the view that financial markets are not simply a reflection of what happens in the real economy. Rather, they can also have substantial impacts on production and consumption of oil, potentially even in the long run. By applying the conception of general equilibrium, this analysis implies that the equilibrium is not stable but instead moves permanently (see, for example, Moore, 2008). The fact that the equilibrium is not only determined by real forces but also by monetary factors radically calls into question the general equilibrium approach in its neoclassical meaning.

The overall analysis conducted so far suggests that the effects of monetary policy are not symmetric. When the central bank raises the short-run interest rate, monetary policy transmission through fundamentals is expected to take place conventionally by entering production cost and affecting investment decisions. The relevant variables take a converse path in the case of an expansionary monetary policy. The financial market aspect, however, is different. First, it is more difficult to speculate for a falling oil price. This would be the symmetric opposite to betting on an increasing price, which is suggested to happen in the course of an expansionary monetary policy. Such speculation might be profitable if a lower oil price is expected. But inducing a lower oil price by financial investment without a change in fundamentals is unlikely for the reasons outlined above. The difference between expansionary and restrictive monetary policy is not that the former leads to higher demand for financial assets while the latter triggers higher supply of assets. The difference is that an expansionary monetary policy raises speculative activity while restrictive policy actions reduce it. Higher interest cost makes investment capital more expensive and profitable investment riskier. Investors' motivation shrinks and money creation for the purpose of financial market investment decreases. Smaller investment capital cannot exert the same demand power and hence financial asset prices tend to fall. The higher the level of interest rates, the less monetary policy transmits through financial markets. The fundamental aspect of transmission remains.

The price effect that originates in the futures market reveals that crude oil has a specific link to inflation. Monetary policy transmission in its conventional understanding occurs through fundamentals and is argued to lead to a quantity effect. Insofar as it leads to rising demand for financial assets, prices of stocks and securities start rising. So-called asset price inflation (see, for instance, Dalziel, 1999–2000; Schwartz, 2002) is not directly linked to the consumer price level and therefore does not lead to a higher inflation rate in the economy. Stock prices can basically fluctuate without corresponding changes in the general price level of the producing economy. Its dual nature makes once more a particular case for crude oil. Its character as an asset and a commodity at once automatically and necessarily transmits price changes caused by financial investment to the spot market. The spot price of oil enters the measured rate of inflation. Even if core inflation is taken instead of overall inflation in order to exclude energy and food prices, the oil price should still affect the general price level, because oil is a widely used raw material input for production (see Cavallo, 2008). Whether this accelerating impact on inflation rates is a demand-pull or a cost-push case depends on the point of view and corresponds to Kalecki's (1987, p. 100) distinction of cost-determined and demand-determined prices: prices of raw material increase when demand increases, owing to inelastic supply in the short run. However, from the perspective of a single company that uses oil as an input, the higher oil price appears as a cost-push that is reflected in the prices of finished goods.

Figure 2.1 has shown how higher demand in the futures market raises the spot price owing to the almost vertical short-run supply curve. Demand in the spot market decreases, but the oil price remains at a higher level and thus raises production cost. In this respect, monetary policy not only leads to higher inflation rates to the extent that higher demand exceeds production capacities in the real economy; it has a second link by affecting commodity prices in financial markets. If the higher oil price induces higher investment and higher capacity utilization and thus raises oil supply in the long run, the inflationary effect might be reversed when the expansionary monetary policy is brought to an end. Basically, the effect may be even greater if the oil price ends up at a lower level than it was before the initial price increase that was caused by monetary policy and speculative activity. The resultant falling oil price not only stabilizes the general price level but should also have a lowering influence on it. The importance of this inflationary impact is an empirical question. Literature usually finds rather limited evidence for higher prices of oil and other commodities to be a source of headline inflation (see, for instance, Cavallo, 2008; Cecchetti & Moessner, 2008). Chen (2009) argues that the pass-through of the oil price into inflation has decreased since the 1980s in most industrial countries. However, the financial market effect on the oil price is a theoretically founded channel concerned with how monetary policy can influence the general price level.

Coming to the end of this chapter, it should once more be taken into account that besides elasticities and market structures, monetary policy is implemented against the background of various additional variables. The effectiveness of monetary policy in its different aspects depends on the state of the business cycle, existing underground oil reserves and the long-run technological trend. On the other hand, the effects that monetary policy has on the market for crude oil might also influence at least some of these background and long-run variables.

2.2.5 Speculation in the Oil Spot Market

A large section of the literature about monetary policy and commodity prices focuses on the kinds of transmission channels that are specific to the oil market (see Anzuini et al., 2013). Frankel mainly developed this approach (Frankel, 1984, 2006, 2014; Frankel & Rose, 2010). It is based on a no-arbitrage condition that contains the interest rate, the price of crude oil, storage cost and convenience yield, and is thus similar to the model of Kaldor (1939) discussed above. The framework takes the form of an overshooting model and thus departs from the assumption that there is an equilibrium price of oil to which the actual price tends to revert in case of deviation. In particular, it is assumed that oil companies have to decide between selling produced oil and holding it as inventories. Sales revenues can be invested at the risk-free interest rate. Alternatively, storing oil brings with it the net benefit of convenience yield and carrying cost as well as the expected price change. The condition of no arbitrage implies that both opportunities yield the same expected return. A change in the interest rate, say, a cut, requires therefore that the oil price be expected to fall for the equation to be satisfied. For this to become possible, the oil price has to shoot up first, such that gradual reversion to the equilibrium price justifies expectations of a falling price. Hence, a falling interest rate leads to a higher oil price. The intuition is given by three transmission channels (Frankel, 2006, pp. 5–8). First, lower returns from investment of sales profits owing to a lower interest rate make it attractive to hold more inventories. Second and analogously, it is relatively more profitable to leave oil under the ground. In this sense, underground reserves are a part of inventories. Third, speculators shift their portfolio from now lower-return bonds to oil futures contracts. While we already incorporated the latter one of these channels by our emphasis on monetary policy transmission through financial markets, the former two add a new aspect to our issue.

Yet, the assumption of the arbitrage condition in question is not without criticism. On the one hand, it may rely on basically logical aspects, but, on the other, it runs the risk of being a theoretical consideration not linked to practice. For instance, it might be true that oil companies hold realized profits in bank deposits or in riskless bonds. However, it might just as well be true, and may be even more likely, that large corporations like those in the oil industry invest their liquidity in other investment alternatives. It is argued that oil sales revenues (the so-called 'petrodollars') flow systematically into stock markets rather than only being invested in government bonds (Varoufakis et al., 2011, p. 326). Moreover, there is more than just one interest rate. Hence, the arbitrage condition requires a preceding assumption about the choice of which specific interest rate to rely on. Producers might also adapt their behaviour to other prices, like the exchange rate. Referring to this finding, it seems that the interest rate is not a key benchmark for oil companies to decide on their inventories. The arbitrage condition that is at the base of the commodity-specific channels is thereby insufficiently verified. Concern about the stability of the arbitrage condition challenges monetary policy transmission through these channels.

Furthermore, it should be mentioned that Frankel's approach not only applies to commodities but likewise to many other sectors where goods stocks are held and where supply and demand are not fully elastic. This would imply that restrictive monetary policy raises inventories and thus contributes to a higher general price level. If this effect held true, the understanding of monetary policy would heavily be shaken, particularly from a monetarist perspective. While such an arbitrage condition is helpful to give some hints about possible 'contango' and 'backwardation' situations in the oil market, it is more critical to use it as a starting point to investigate monetary policy transmission.

Despite the probable weakness of the arbitrage condition, this approach raises a question that deserves to be discussed. It is about whether or not oil companies actively alter inventories in response to monetary policy. In other words, do oil producers behave speculatively by raising or lowering their volume of stocks? Our hitherto conducted investigation has shown that as long as the price elasticity of oil demand is larger than zero, inventories inevitably have to increase in order that speculation can have an effect on the oil price. Stock accumulation in this case is driven by increasing demand for futures contracts in the futures market. Hence, speculative activity in the futures market is the force that causes oil stocks to augment. Inventory accumulation is a reaction to a higher oil price. Frankel's assumption of no arbitrage and the suggestion that oil producers may themselves behave speculatively turn causality around: it is argued that inventory accumulation is the reason for a higher oil price, because the behaviour of oil companies produces scarcity in the spot market.

Speculation rooted in the spot market is different from speculative activity in the form of financial investment. Speculation in the futures market is demand-determined. Increasing inventories and decreasing spot demand are the result of overcompensating demand for oil futures contracts. Speculation in the spot market is supply-driven in the sense that the supply side of the crude oil market (oil companies) is the origin of the price change. They constrain oil supply by increasing stocks. Financial speculation in the futures market appears as an exogenous force to the spot market such that the price increase is not *a priori* reversed by reactions in the spot market. Speculative behaviour of oil producers is an endogenous factor. Supply shortening has – among other impacts – an effect on demand that redounds on supply owing to its endogeneity. Raising the oil price by lowering supply might reduce demand to such an extent that profits of oil companies decline. No firm would probably be willing to do that. In the futures market, speculative demand pushes up the price. But since the number of contracts is not limited and as an increasing price does not lower demand but raises it further because of investors' profit expectations, there are no direct correction factors that would lead the futures price automatically back to its initial level. It is only over time that speculative price hikes in financial markets require corrections owing to their contradiction with the real economy. In the oil spot market, however, speculation has direct constraints owing to conventional supply and demand feedback mechanisms.

The likelihood that oil companies behave speculatively is determined by a number of factors. First, price elasticity of demand must not be too large. A small elasticity estimate allows a price increase to persist without being fully reversed by lower demand. Second, competition in the oil spot market cannot be perfect. There must be a certain degree of monopoly. Otherwise, the speculative supply cuts of one oil company are compensated by another. Perfect competition would imply full-capacity utilization (see, for example, Kalecki, 1987, pp. 71–82). Third, in the case of a high degree of competition, the price elasticity of supply should be sufficiently low. This means that competition takes a certain time to evolve such that supply cuts by some companies are not evened up immediately by others. A fourth factor might be the amount of existing reserves. In expectation of exhausted global oil reserves in the near future, oil companies should tend to raise oil inventories. Inducing a price increase by the augmentation of inventories is easier to afford when oil producers expect that the oil price will rise soon, in any case. Competition in the market may penalize speculative behaviour by a loss in market shares in the short run. But strategic behaviour will bring longer-run profits when oil reserves become scarcer.

The more these factors apply to reality, the more the oil spot market is able to affect the oil price by speculative manipulation of inventories. If they do not apply, apart from some certainly existing time lags, a higher oil price – in the absence of speculation in the futures market – leads to lower inventories, because the only thing non-speculative oil producers aim at is the satisfaction of a growing demand for oil. As usual, with respect to the issue of speculation, these factors probably apply to some – though limited – degree but they do not allow the manipulation of the oil price in the long run. If accumulated inventories are not sold, they do not contribute to oil companies' cash profits. To realize profits, sufficient effective demand requires a sufficiently low price level. Hence, the oil price cannot stay at too high a level for too long.

Dvir and Rogoff (2014) show that there is a long-run cointegrating relationship between supply, demand and inventories. They argue that under the condition that supply is inflexible, higher demand leads to higher inventories. This means that producers and traders behave speculatively as they expect that this will raise the price further. Supply inflexibility is suggested to have lasted since 1973 (ibid., p. 114). Even though this is only a long-run result without any indication about the direction of causality, its intuition can analogously be applied to our short-run considerations.

The argument that a higher oil price is correlated with higher inventories if the former is driven by speculation is well accepted. While we claim that the causality goes from the oil price to inventories owing to financial speculation in the futures market, we shall not fully reject that an additional impact of speculation might be exerted by causality from stocks to price. However, we suggest financial speculation to be stronger and longer-lasting, because the futures market does not face the same constraints that the spot market does.

We argue that the interaction between the financial market and the fundamental aspects of expansive monetary policy transmission to the oil market induces oil producers to raise supply: a higher oil price originating in futures market speculation raises the profit expectations of oil companies. Production should therefore grow. This may appear as a contradiction to the argument of speculation in the spot market, where oil companies reduce rather than raise oil supply, yet there is heterogeneity in the oil market and producers may have different strategies according to different time horizons. Heterogeneity means that some companies might speculate while others may behave simply competitively. Moreover, a company may reduce oil supply at the short horizon but simultaneously raise investment to be able to increase production in the future. Hence, both possible outcomes are compatible as we suggest speculation in the spot market to be effective only in the short run.

NOTES

1. In this respect, see Clower's (1967, p. 5) famous dictum: 'A commodity is regarded as money for our purposes if and only if it can be traded directly for all other commodities in the economy. [. . .] money buys goods and goods buy money; but goods do not buy goods'. It has served as a foundation for many influential contributions to neoclassical monetary theory, as for instance, Kiyotaki and Wright (1989).
2. See Romer (2000) for an elaboration, and Lavoie (2006b) and Rogers (2006) for a critique of the New Consensus on monetary policy.

3. Cencini (2003a, pp. 313–314), as a representative of the theory of money emissions, even rejects the notion of a 'stock of money', since he argues that money is immaterial. We agree but may nevertheless use the notion in some places for convenience where it is helpful to explain the endogenous character of money.

4. The fundamental difference between the theory of the monetary circuit and the theory of money emissions consists of the meaning of money. The former considers money as a stock in circulation, that is, being in a flow. The latter argues money itself to *be* the flow (Rossi, 2009b, p. 39).

5. Rossi (2001, pp. 139–145) argues that this kind of inflation is not cumulative over time. Excessive credit creation leading to demand exceeding production capacities and hence rising prices is reversed once credits have to be paid back. Instead, structural and cumulative inflation arises from the working of the contemporary monetary and banking system and is not caused by monetary policy (Gnos, 2007; Rossi, 2001, pp. 145–153, 160–169). Here, we limit the investigation to changes in the general price level that are potentially caused by monetary policy.

6. For detailed elaboration, see for instance Frank (2008) and Jehle and Reny (2011).

7. For a review of the theoretical and empirical literature on herding in financial markets, see for instance Bikhchandani and Sharma (2001).

8. It may as well be the case that the bank's deposits exceed its loans such that it possesses a net deposit with the central bank's balance sheet.

9. Note that Table 2.2 as well as Table 2.3 show the simplified economy from the production instead of the income side. The income side consists of bank deposits that are claims to purchase total output.

10. Note that we do not use the term of liquidity preference as it is used by Keynes (1936/1997, p. 241) against the background of a vertical money supply curve. Rather, we apply it as a concept integrated into the horizontalist perspective of money (see, for instance, Erturk, 2006, p. 466; Kaldor, 1985, p. 9; for an overview of the discussion, see Cardim de Carvalho, 2013).

11. Interestingly, Estrella (2002) finds that securitization of assets has made the pass-through of the monetary policy target rate of interest to market rates stronger. On the other hand, policy effects on economic activity seem to have decreased. This may strengthen the view that financial markets, whose weight has increased in the course of financialization, may evolve remarkably independently of economic fundamentals.

12. For an analysis of how capitalism produces endogenous crises, see for instance Minsky (1982, 1994).

13. For a general critique of neoclassical trouble with time, see for instance Varoufakis et al. (2011, pp. 156–158).

14. To be exact, capacity utilization is not necessarily permanently equal to unity in neoclassical models. But distortions in full utilization are due to exogenous shocks that affect the real business cycle and nominal rigidities that delay adjustment to the general equilibrium state (see, for instance, Greenwood et al., 1988; Svensson, 1986). Hence, once the business cycle is modelled and since the markets always clear by model construction, there is no way that inventories can be accumulated nor capacity utilization varied by entrepreneurial decision, owing to effects like speculation in financial markets.

15. The BRICS notion contains Brazil, Russia, India, China and South Africa.

16. Even though we still agree with the argument that monetary and real terms cannot be divided in a monetary economy (Cencini, 2003a, pp. 303–304), the comparison of the oil price to the general price level is useful. In fact, it is an *ex post* comparison of two monetary terms and hence is in line with our monetary analysis. In contrast to neoclassical economics, we do not use the 'real' price of oil to make predictions or to measure causalities. Indeed, in economic reality, only the nominal oil price is visible.

17. Concerning the issue of portfolio diversification, many studies besides that of Gorton and Rouwenhorst (2006) investigate the correlation between equity returns and the oil price. Both negative and positive correlations are found (see, for instance, Kolodziej et al., 2014; Lee & Chiou, 2011; Miller & Ratti, 2009; Tang & Xiong, 2011). In the

studies, correlation is shown to change over time. While a negative correlation confirms crude oil futures to be effective diversification opportunities in addition to stocks, positive correlation is interpreted as a sign of increasing financialization that links equity and commodity markets. We do not judge these results here. It is sufficient to mention that a change in the correlation is not a theoretical contradiction. In fact, the positive sign may just be linked to the negative one: Investors might exploit the negative correlation for portfolio diversification. Demand for assets thereby exerted may produce a positive correlation. Fluctuations in the empirically measured relationship between commodity and equity prices thus should be the rule.

18. This may well occur but is based on strategic behaviour in monopolistic markets rather than triggered by monetary policy (see, for instance, Varoufakis et al., 2011, pp. 345–346).

19. It may seem a contradiction to the idea of an accommodative supply of money, since demand is not expected to have an influence on the 'price' of money (that is, in this case, the value of the domestic currency in terms of the foreign currency, that is, the exchange rate). For an explanation of exchange rate changes against the background of endogenous money, see Cencini (2000, pp. 11–14). For the purpose at hand, the conclusion about interest effects on the exchange rate does not change.

20. In the United States, for instance, there is a large number of small oil producers. However, they make up only a small fraction of the total output (Meyer, 2014).

21. Minimum reserves do not contradict the horizontalist argument of accommodative central bank behaviour. Reserves are supplied according to demand (see Rochon & Rossi, 2011). However, interest rates on minimum reserves involve higher costs for banks.

22. Owing to financialization, it becomes even more likely that production companies do not supply financial assets by the emission of stocks and bonds, but become more and more demanders of financial assets by participating in financial investment (see, for instance, Lazonick, 2012).

23. Remember the anecdotal citations in the introduction.

PART II

Monetary policy and crude oil in the real world

3.   US monetary policy and the global crude oil market

In this part, we connect the assessed theoretical and partially abstract results to the current practice of central banking. Furthermore, we elaborate upon the crude oil market in its geographic and temporal integration and present the ins and outs of the practice of crude oil pricing. Next, we explain how crude oil is connected to its closest substitutes in the energy market. And finally, we investigate how the monetary policy of a single country can affect the global crude oil market.

3.1   US MONETARY POLICY IN THE TWENTY-FIRST CENTURY

Until now, we represented monetary policy by simple manipulation of the short-run interest level. Yet, practical implementation is more sophisticated. Moreover, and crucially for empirical analysis, the so-called 'unconventional' monetary policy should be emphasized.[1] It works partially analogously to conventional policy but features as well specific mechanisms and specific transmission channels. They are treated in the following section to provide a proper understanding of monetary policy practice and how it is integrated into the hitherto framework.

3.1.1   Basic Mechanisms of Monetary Policy Implementation

The United States Fed pursues three main long-run goals of monetary policy, that is, 'maximum employment, stable prices, and moderate long-term interest rates' (Federal Reserve System, 2014b). While the goal of price stability is quantified by a targeted inflation rate of 2 per cent, the Fed does not locate fixed long-run values of employment and interest rates, as they depend on respective economic circumstances. In the short run, daily monetary policy actions have the target of the federal funds rate at the centre. The fed funds rate is the interest rate at which financial institutions with access to central bank reserves – which usually are commercial banks – lend those reserves to one another. This kind of interest rate targeting corresponds well with the approach of endogenous money, which argues that the interest rate rather than monetary aggregates is the exogenous variable.

The federal funds rate is the result of supply and demand in the market for reserves. It consists of reserves from the central bank, which are generally referred to as 'borrowed reserves' (see, for instance, Ennis & Keister, 2008, p. 241). Borrowed reserves are granted to depository institutions at the discount rate. Reserves that are exchanged in the interbank market are 'non-borrowed reserves'. It is changes in reserves or in the discount rate that allow the central bank to reach the targeted level of the federal funds rate in particular and thereby influence interest rates across the whole economy. Arbitrage activity connects the target rate and market rates of interest. The transmission does not lead to exact equalization of different rates but only determines the direction of changes (see Atesoglu, 2003–04; Payne, 2006–07). Market rates are, naturally, further influenced by exchange rates and financial market factors.

In practice, there are three tools that the US Federal Reserve can use to realize the target funds rate (for details about practical implementation, see, for instance, Krieger, 2002, pp. 73–74). First, it determines the quantity of open-market operations. By means of these so-called repurchase agreements (repos), the central bank lends reserves to depository institutions under the agreement to pay them back within a time span of a few days up to several weeks during normal times. Open-market operations are backed by securities, which serve as collateral and are offered by the borrowing institutions. The second instrument is the setting of the discount rate. It transmits to the interbank market and hence has an influence on the federal funds rate. Nevertheless, it is – taken alone – not sufficient to guarantee the achievement of the target funds rate level, as is explained below. Third, banks are required to hold at least a minimum level of reserves. They are measured as a fraction of total transaction deposits of banks. Reserve requirements are on the one hand suggested to reduce risk in the banking sector, because they lower the probability of a specific bank getting into liquidity shortage when depositors raise demand and take off their balance in an unexpected moment. On the other hand, the use of this tool is suggested to control the banks' issuing of loans to the public through the restriction of the minimum share of reserves in balance sheets. Reserve requirements are clearly subject to the exogenous-money approach, as they rely on the concept of the monetary base and the money multiplier. Thus, they are naturally often an issue of criticism. Rochon and Rossi (2011) argue that reserve requirements are an ineffective instrument to control the volume of money creation because central banks behave in an accommodative way.

In a horizontalist framework, they would otherwise provoke a liquidity crisis.

Institutionally, the US Fed consists of twelve regional Federal Reserve Banks that coordinate the conduct of monetary policy. The Federal Open Market Committee (FOMC) meets regularly and decides about open-market operations. A faction of the FOMC, the Board of Governors of the Federal Reserve, is responsible for the setting of the discount rate. Finally, the FOMC determines the target level of the federal funds rate. The decisions are communicated to the public.

3.1.2 Conventional Monetary Policy

The conduct of conventional monetary policy implementation in the United States in the twenty-first century exhibits a distinct change in the year 2008 as explained, for example, by Lavoie (2014, pp. 221–223). Prior to 2008, the FOMC used to set the federal funds rate target somewhere between zero and the discount rate. Arbitrage between borrowed and non-borrowed reserves tends to equalize the funds rate and the discount rate. But equality between the rates is not effectively reached. The federal funds rate is the endogenous result of supply of, and demand for, borrowed as well as non-borrowed reserves. The discount rate, contrastingly, is exogenous and sets the condition for borrowed reserves. The existing amount of reserves is reflected in the federal funds rate. The central bank can lead it to the target level by setting the discount rate, that is, the interest rate for additional reserves, above the target level. This prevents borrowed reserves from growing further. In this way, the federal funds rate is consolidated at its level. Even though arbitrage does not result in the equivalence of interest rates, it is responsible for the transmission of changes in the exogenous rate to the endogenous rate. This system reveals a basic problem: the central bank aims at realizing the interest rate target. For this purpose, it has to manipulate the supply of reserves by setting the discount rate. At the same time, it is effectively obligated in practice to accommodate changes in reserve demand so that the financial system is not jeopardized. Hence, there is a trade-off that complicates the successful achievement of the federal funds rate target. In contrast to the system after 2008, the central bank faces more difficulty to reach the target rate as it has only one benchmark to manipulate, that is, the discount rate. This has changed since then.

In the course of the financial crisis in the second half of 2008, the Fed modified its monetary policy implementation and adopted a corridor system (Lavoie, 2014, pp. 223–225). The first crucial difference is that the deposit facility rate, to wit, the rate at which banks can deposit their reserves at the central bank, became positive while it had been zero before. Logically, it is always lower than the discount rate. Arbitrage prevents the climbing of the federal funds rate above the discount rate. In the same way, if the federal funds rate were lower than the deposit rate, the interbank market would break down, because all banks would deposit their excess reserves at the central bank, which would guarantee them a higher profit. The discount rate and the deposit facility rate are therefore the upper and lower bound of the target rate, respectively.

The second change in 2008 was the choice of the federal funds rate target. There are basically three different systems: the floor, the ceiling, and the symmetric system (Lavoie, 2014, pp. 223–225). They set the target rate of interest at the lower bound, the upper bound or symmetrically in between. The US Federal Reserve adopted the floor system. It means that the target rate and the deposit facility rate are equal. The system implies a large supply of reserves to bring down the federal funds rate sufficiently. This brings the advantage in that the central bank is more flexible. Before, reserves were demand-driven to the extent that the federal funds rate target was not violated. Under the floor system, even this last constraint on reserves is removed. Given the large supply of reserves that equalizes the federal funds rate and the deposit rate, reserves can be acquired in the interbank market without any loss, because they can be deposited at the same rate with the central bank in case of excess reserves. Banks do not have to optimize reserves to a minimum anymore. Hence, monetary policy can be made without facing any trade-off between the target rate and stabilization purpose when the financial system is in need of reserves. Open-market operations and reserve policy are now two completely independent policy instruments (see, for instance, Goodfriend, 2002, p. 6). One might argue that the central bank can now control the supply of reserves as well as control the interbank rate (see Lavoie, 2010, p. 18). However, given that the supply of reserves must in any case be sufficiently large to keep the target rate at the deposit rate, the quantity of reserves that finally comes into existence is determined by the demand for reserves. No more reserves can exist than depository institutions want to borrow. The need for reserves is itself dependent on credit demand from investors in the economic system. Under a floor system, banks might borrow quite large reserves and deposit them at the central bank. Such reserves exist but they are, in fact, unemployed. Their employment is again driven by demand. In this sense, it is reasonable to conclude that the adoption of the floor system has completed the demand-determined nature of endogenous money.

3.1.3 Unconventional Monetary Policy

The outbreak of the financial crisis in 2007 has led to fundamental changes in how central banks make monetary policy. In particular, it was in 2007 when the US Federal Reserve started taking unconventional measures to provide financial markets with liquidity. In general, unconventional monetary policy is perceived as offering open-market operations at extraordinarily low discount rates or as large asset purchases by the central bank.

Quantitative easing is the most common term for asset purchases. It is the most important tool of unconventional policy and will subsequently be focused upon. Depending on the type of assets, Borio and Disyatat (2009, pp. 7–8) distinguish exchange rate policy, quasi-debt management, credit policy and bank reserves policy. The former three types refer to the purchase of foreign currency, government bonds and private sector assets, respectively. Bank reserves policy corresponds to a policy stance where the monetary authority targets a specific amount of reserves. The US Federal Reserve unfolded its activity in quasi-debt management and credit policy.

Asset purchases can be neutralized by counteracting actions (like repos). This is what the Fed did in 2007 and partially in 2008 (Lavoie, 2010, p. 3). In this case, the balance sheet of the central bank remains constant in its length. What took place after quantitative easing without neutralization has lengthened the central bank balance sheet to a hitherto unseen volume. The federal funds rate approximated zero in December 2008 (ibid., p. 8). Expansive monetary policy in its conventional form of manipulating the federal funds rate found its limits and became ineffective. Hence, the change to quantitative easing was not a free choice but rather a necessity from the perspective of the central bankers.

Unconventional monetary policy is often understood in the public domain as a flooding of the economy by huge amounts of free money. As McLeay et al. (2014, pp. 21, 24–25) explain, central bank purchases of financial assets, say, government bonds, compensate the original bond holder by an equal amount in the form of money that the latter holds as a bank deposit. The thereby increased liabilities of the commercial bank are rebalanced by a corresponding increase in central bank reserves that the bank holds. Reserves enter the central bank balance sheet on the liabilities side and thus equate it as its assets side has grown owing to government bonds purchases. Hence, while the balance sheet of the original asset holder remains constant, the balance sheets of the commercial bank and the central bank both increase by the sum of the assets' market value. Net wealth has not increased for anybody; quantitative easing creates no free money. The only change in this respect may occur by an increase of asset prices that takes place when the central bank exerts demand for a limited number of assets.

Central bank reserves used to be demand-determined. Under the unconventional condition of a federal funds rate close to zero, however, there is no further cut in the rate that would trigger higher demand for reserves. The central bank tries nevertheless to raise the amount of reserves in order to relax tight liquidity conditions in the interbank market and in the rest of the economy. Even though reserve demand stagnates, quantitative easing raises the amount of reserves. Under these extraordinary circumstances, reserves are supply-driven (Lavoie, 2014, p. 226). Since depository institutions do not demand additional reserves, the monetary authority injects them into the economy by circumvention of the banking sector and direct asset purchases from the non-bank sector (Borio & Disyatat, 2009, p. 16).

In this sense, quantitative easing serves as an important tool to the central bank, since it can in principle expand the balance sheet of the whole banking system by injecting reserves into the economy independently of the commercial banks' will. However, there is also a specific limitation with regard to quantitative easing that does not apply to conventional monetary policy: the monetary authority has the monopoly in setting the short-run interest rate, which transmits to the economy through various channels. Balance sheet manipulation, in contrast, can basically be made by any market participant. Even though the central bank has unlimited means and is a powerful agent, balance sheet policy is not its exclusive property. Control of interest rates and price variables by means of quantitative easing is more difficult than in the case of conventional monetary policy (Borio & Disyatat, 2009, p. 14).

Yet, the fact that reserves may become supply-determined in the case of a floor system and in the presence of expansive monetary policy does not imply that the amount of money is also driven by supply. Reserves themselves do not necessarily increase proportionally to the supply intended by the central bank. They may be traded on the interbank market and be used to reduce overdraft reserves by some commercial banks (Lavoie, 2014, p. 228). But it can at least be said that once banks do not aim at optimizing reserve holdings anymore, they just cannot get rid of reserves, so that the central bank has many options for influencing the reserve market. In the case of broad money, things are less easy to control. Asset holders who sell their assets to the central bank might use the newly created money in their possession to repay debt. Hence, while reserves increase by a specific amount, the quantity of money may rise only by a fraction of it (ibid., p. 228). Money thus remains demand-determined even in the presence of unconventional monetary policy.

Quantitative easing or, more broadly, unconventional monetary policy may be seen just as another form of expansive policy since they share the same purposes in what concerns inflation, output, employment and other potential target variables. Yet, strategies of conduct differ in important aspects. Specifically, unconventional monetary policy by definition only takes place in unconventional circumstances. In recent years, quantitative easing suggested itself on the one hand because the federal funds rate reached the zero lower bound. On the other hand, the transmission mechanism of monetary policy seemed to be broken: market interest rates diverged from the rates targeted by the central bank (Joyce et al., 2012, p. F276; Pollin, 2012, pp. 67–68). Risks in the market were judged as too high for policy stimulation to have any further influence. Unconventional monetary policy is therefore often seen as a tool to re-establish the transmission mechanism in order for conventional policy to work again (Joyce et al., 2012, p. F272). Quantitative easing takes place against the background of a two-stage transmission of monetary policy. The first step aims at reconnecting the target interest rate and market rates. The second stage consists in the transmission of altered interest rate levels to economic activity. Since the second phase is largely seen as analogous to the transmission of conventional monetary policy, which thus does not need to be emphasized further, studies about quantitative easing usually concentrate on the first phase. This does not mean, however, that unconventional monetary policy does not have any effect on real economic variables, the price level or financial markets (see, for instance, Kapetanios et al., 2012). These effects can be direct as well as indirect and will be important for our specific issue of the oil market. But the outstanding features of unconventional monetary policy require the enlightening of the first stage of its transmission.

Quantitative easing is argued to exert effects mainly on interest rates and asset prices. Asset purchases should raise their prices. Higher prices of securities directly translate into lower interest rates. This helps to lower the general level of market rates of interest and to lead it closer to the target rate, which is zero in the present case of examination. Research literature has elaborated specific transmission channels for the first stage, that is, from policy action to market interest rates. As in the case of transmission to the real economy, first-stage transmission channels are based on different arguments and thus are not perfectly complementary but, rather, overlapping. According to authors, as those referred to in the following, they differ in names and numbers. The following listing includes the transmission channels generally referred to in the literature.

The first channel is the often-mentioned portfolio balance channel, portfolio substitution channel, or scarcity channel (see D'Amico et al., 2012, pp. F424–F425; Joyce et al., 2012, pp. F277–F278). It is based on the realistic assumption that assets in private balance sheets are not perfectly substitutable. By purchasing illiquid assets against newly created money, private sector balance sheets are relaxed, which is mirrored in the altered behaviour of the owners who are ready to accept lower interest rates on their other assets. Given that investors have a preference for an asset class of specific duration and degree of liquidity, purchases by the central bank affect relative scarcities of asset classes and thereby raise the price of focused assets, because they are not (perfectly) substitutable. The price increase of securities lowers their return and thus supports the reduction of the interest rate level (D'Amico et al., 2012, p. F425). Portfolio rebalancing of investors spreads the effects to other interest rates.

A second way of transmission is the expectations or signalling channel (Borio & Disyatat, 2009, p. 13; D'Amico et al., 2012, p. F424). As agents build expectations that are, among other things, based on what the monetary authority communicates, they may anticipate future policy conduct. Asset purchase announcements by the central bank convince investors that interest rates will stay low or at zero for a long time. This lowers long-run rates of interest and hence contributes to the re-establishment of the relationship between short-term and long-term interest rates.

Furthermore, D'Amico et al. (2012, pp. F425–F426) suggest a duration channel. The interest rate of long-term assets is composed of the short-run rate plus a term premium that reflects the higher failure risk during the longer term. Higher demand for such long-term assets exerted by the central bank raises their price and liquidity, and thus lowers their risk.

Risk does not only have a time dimension but also materializes between assets of different classes. Similar to the purchase of long-term securities, central bank purchases of riskier assets *ceteris paribus* raise their price and lower the yield relative to safer assets. Thus, not only the risk-free share of the interest rate falls but also the premium assigned to riskier securities (Joyce et al., 2012, p. F279). This channel may generally be denoted as a risk premium channel.

Borio and Disyatat (2009, p. 14) suggest not only a higher demand for riskier assets due to the portfolio balance channel but also identify a distinct risk-taking channel. Investors face a lower interest rate level owing to quantitative easing. This incites them to search for yield and hence invest in riskier assets. This has the corresponding effects of higher prices of other assets, be they securities, stocks or commodities.

While any channel is likely to affect the degree of liquidity of the asset classes concerned, quantitative easing may have the basic impact of providing any amount of liquidity in an acutely dried-up market. Gagnon et al. (2011, p. 8) therefore mention a liquidity channel.

These transmission channels exhibit the outstanding features of unconventional monetary policy. The two stages of transmission are distinguished in favour of a precise analysis. In practice, however, it may be hard to identify them separately. For example, Krishnamurthy and Vissing-Jorgensen (2011, pp. 223–224, 240–243) define an inflation channel that specifically relates to quantitative easing. It works by way of an increase in expected inflation due to expansive monetary policy. Inflation expectations then enter market interest rates by lowering real long-run rates. Should this inflation channel really exist with regard to unconventional monetary policy, then it should also exist in the context of conventional policy. Changed inflation expectations in the course of interest rate cuts by the central bank should have their additional effects on market rates of interest. Yet, the neoclassical school of thought suggests exactly the opposite. It argues that it is inflation as well as inflation expectations that render an active expansionary monetary policy rather ineffective (Friedman, 1968, pp. 5–7). We know that monetary policy does not automatically raise the inflation rate by augmenting reserves. On the other hand, market participants may nevertheless potentially build their expectations on grounds that are not necessarily appropriate. Thus, neoclassical theory seems to be somewhat contradicting with regard to the inflation channel, while our alternative view raises fundamental doubt about the existence of that channel.

Be that as it may, we do not explicitly list such channels since they involve second-stage transmission from market rates of interest to the real economy: they apply if the first-stage channels have been successful. Yet, the two phases are hard to distinguish in practice. Moreover, it seems appropriate to say that the effects of unconventional monetary policy eventually materialize through the same transmission channels as in the case of conventional policy, that is, the interest rate channel, exchange rate channel, Tobin's q, wealth effect and credit channels.

The impact of quantitative easing on financial markets should be more pronounced than that of conventional monetary policy owing to a structural difference in the character of implementation. By manipulating the interest rate level in the course of conventional monetary policy, the central bank sets the general framework for the economy. A cut in the interest rate affects the real economy and financial markets comparatively more or less, depending on endogenous variables like effective demand, expectations, liquidity preference and others. It prefers neither real nor financial investment *a priori*. Quantitative easing, in contrast, has its direct focus on financial markets. The monetary authority itself takes on the role of a financial investor; not just a common investor, but also the one who dominates all others. It directly exerts demand power in financial markets, while conventional policy only sets a precondition for it by cutting the interest rate. We argue that conventional expansive monetary policy leads to larger financial investment through many mechanisms that induce investors to do so. In the case of quantitative easing, larger financial investment is simply a fact. This finding is far from postulating a linear relationship between unconventional monetary policy and demand for financial assets. In a crisis time, it may be that investors are just glad to get rid of illiquid assets and use the monetary sales return to repay debt. Thus, we only apply this argument to the concrete action of the central bank. Obviously, however, the entrance of the central bank into financial markets improves expectations of speculative investors and hence is likely to trigger further financial investment. This is again analogous to the effects of conventional policy.

Owing to the omnipresence of uncertainty in the economy, there is an important limitation to quantitative easing. By means of the signalling channel, the monetary authority tries to convince the public of its policy strategy. The commitment to take extraordinary measures to stimulate the economy may make investors more optimistic. In specific uneasy situations, however, market participants may also interpret unconventional monetary policy measures as a sign that the state of the economy is worse than expected. Hence, the announcement of new measures might trigger a strong movement towards safe bonds and away from risky assets (Neely, 2011, p. 23). Long-run lending rates of interest thereby may increase even further. Reasonably, longer-lasting asset purchases by the central bank are likely to overcome such counteracting capital flows.

The neoclassical school of economic thought might suggest that quantitative easing is a guarantee to raise inflation rates. Even though usually considered as harmful, an inflationary effect can be welcome in an environment of potential deflation. Kapetanios et al. (2012, p. F316) argue that quantitative easing in the United Kingdom aimed at 'inject[ing] a large monetary stimulus into the economy, to boost nominal expenditure and thereby increase domestic inflation sufficiently to meet the inflation target'. There is the assumption that a higher amount of money translates into a proportional increase in the rate of inflation measured in the goods market. We reject this assumption in light of the endogenous-money perspective, because an increase in reserves does not affect the general price level if additional demand flows into financial markets only. Quantitative easing is only inflationary in the goods market, if its transmission leads, first, to higher lending to the real economy instead of only financial markets. Second, higher lending has to lead to effective demand that increases production capacities. If these conditions are not fulfilled, any inflationary effects are due to price increases in financial markets that transmit to fundamentals. We have treated this argument above.

3.2 THE GLOBALIZED PRICING SYSTEM OF THE CRUDE OIL MARKET

In the hitherto analysis, the oil market has been considered as a model market where there is a spot price and a futures price, both uniquely identified by market supply and demand forces. On this foundation, we have criticized neoclassical theory and, specifically, the efficient markets hypothesis according to which futures prices are a simple reflection of spot market conditions. As we will show in this section, the global oil market does not only consist of single prices in the spot and in the futures market, respectively. It is also a complex set of interlinked prices and production quantities. We will point out that the reality of the crude oil market does not contradict our investigation results but, in contrast, reveals many features that challenge the neoclassical assumptions of efficient markets even more. Fattouh (2011) provides a detailed study of how oil prices materialize worldwide to which we largely refer now.

There exist at least as many different types of crude oil as there are oil-producing countries and regions in the world. The most important among them are WTI in the United States, Brent in the North Sea, and Dubai and Oman as representatives in the Gulf region. They serve as a benchmark on which other crude oils rely. Concretely, the benchmarks set the price level and the dependent types of oil set the price differential, which is largely made up of differences in quality, that is, density and sulfur content. The differentials can, however, change temporarily as supply and demand conditions of a certain type of crude oil alter relative to other types (Fattouh, 2011, p. 21).

Neoclassical economic theory assumes an efficient market mechanism that sets the price of a good where supply meets demand. Yet, in reality, things are less clear. Supply and demand are not curves given by nature and the crossing point is not unique. Rather, there are many individual suppliers and demanders who agree on the transfer of goods at a bargained price. Since many such deals take place, many different prices exist. The market mechanism leads in fact to an indeterminacy of price (see, for instance, Bénicourt & Guerrien, 2008, pp. 27–28). This exactly applies to the global market for crude oil, where prices of trades differ across different types of oil as well as within them. The single market price does not fall from heaven but has to be assessed by calculation based on individual deals. This is the task of price reporting agencies (see Argus, 2015; Platts, 2015). The calculation methods are not the same for all agencies. They consider a time frame of distinct length and take into account the deals that take place within this period. Some also include undone deals, that is, the corresponding bids and offers (Fattouh, 2011, p. 31). A method where deal prices are weighted and averaged over several hours a day faces the problem that the price resulting at the end of the day is biased since it contains many past prices rather than merely the actual price at the close. In contrast, short frames at the end of the day do not fear great lags but may suffer insufficient liquidity such that few great traders may lead to distortions in the price result (ibid., p. 32). Different methods hence imply different price assessments for one and the same type of crude oil.

All in all, price reporting agencies try to cover market developments as carefully as possible. But there is also an impact in the reverse direction. The particular pricing system chosen influences the market and participants in their buying or selling decisions, which may have an additional effect on the price (Fattouh, 2011, p. 30).

The need of price assessment shows that it is doubtful whether a true market price ever exists, since the market consists of individual deals such that a single price is quite hypothetic. Once allowing for the abstract assumption that there is a unique equilibrium price, the process of oil price calculation shows that there are many ways for the assessed price result to deviate from the invisible true price. We agree that supply and demand forces do, of course, have their respective impact on prices and quantities. But exact crossing points in a mathematical sense do not reflect reality. This gives rise to the conclusion that if the spot price of crude oil is allowed to deviate from its own theoretical market price, then the futures price should at least also be allowed to deviate from it. This argument supports the view that the efficient markets hypothesis fails in the face of economic reality. The pricing systems of the dominant benchmark oil prices show that there is even more doubt in considering the futures price as a simple reflection of the spot market.

The WTI will be at the centre of the forthcoming analysis. It is not the only dominant benchmark but has nevertheless some outstanding characteristics. Almost all oil imports to the United States are priced with reference to WTI. That is, they are not all of WTI type but they take it as a benchmark and set a specific price differential that covers qualitative differences of crude oils. The United States consumes a narrow quarter of total worldwide oil consumption (Fattouh, 2011, p. 52). Moreover, and crucially, the Light Sweet Crude Oil Futures Contract is one of the largest traded commodity futures contracts and is based on WTI. Its trading volumes on the New York Mercantile Exchange (NYMEX) have grown exorbitantly in the decade from 2000 to 2010 (ibid., pp. 54–55).

In practice, identification of the WTI spot price is done by taking the level of the WTI futures price as the starting point. The spot price comes into play by the posting plus (P-Plus) and the differential to NYMEX Calendar Monthly Average (CMA) markets, which are both spot markets.

Both use the futures price as a reference, so that what is effectively traded in these markets is the differential to the futures price (Fattouh, 2011, p. 58). Without going too far into details, this price assessment reveals two important aspects. First, the futures price is the basis of calculation, that is, it sets the price level. The spot market only sets the difference. Second, the spot price contains an element of forwardness (ibid., p. 20). Futures prices are real-time prices since they suffer hardly any time delay. In contrast, physical deliveries usually do not take place immediately. Time inconsistency is likely to arise for the reason that the spot price is fixed for a quantity that is only to be delivered in the time to come. Thus, it is probably not a perfect reflection of supply and demand conditions in the spot market at the time of delivery. This means that the spot price may already be different from the equilibrium value in the sense of neoclassical theory before we start any examination of the issue.

It is quite hard to derive clear causalities from these findings. The price level is provided by the futures price but this does not necessarily imply that the futures price is the principal determinant variable of the spot price. Expectations of investors can react to the evolution of fundamentals by entering the futures price in such a way that it changes simultaneously with the spot price. This has been shown in Figure 2.1. It appears nevertheless as a criticism of the assumption that the futures market is fully determined by developments in the spot market. The fact that the futures price is given while the spot price has to be assessed by reference to the former leaves additional room open for changes in the futures market to materialize in the spot market.

It is this complex network of price levels and differentials that brings the distinct types of crude oil together and forms a world market for them. Regional distances and different methods in price assessments may limit arbitrage between crude oil benchmarks. However, worldwide oil transportation connects regions and requires benchmarks to follow the same long-run paths. A comparison between the two main benchmarks exhibits this fact quite convincingly. Figure 3.1 exhibits the WTI and Brent spot price evolution of past years. They used to move closely together. The only lasting differences are found between 2011 and 2013, when there was a WTI discount of up to 25 US dollars. This gap is mainly explained by supply constraints due to limited pipeline infrastructure to the US Gulf Coast. Higher cost of alternative transportation required a lower WTI price. This shortage was eventually overcome by higher capacities (EIA, 2013). The correlation between both prices from 2000 to 2014 at daily frequency is 0.98. This allows us to assume in the following that the oil market is globally integrated. It is thus fair to refer to WTI, thereby covering the world market. We should, however, be aware of the potential limitations

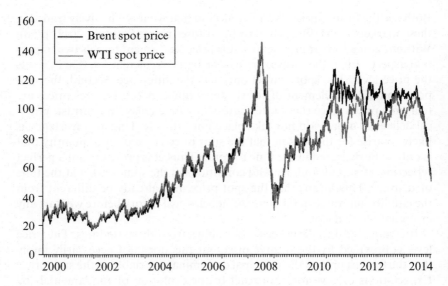

Source: Energy Information Administration (2015a). Petroleum and other Liquids.

Figure 3.1   WTI and Brent spot prices, 2000–14, in US dollars

and noise that this approximation can bring to our results. Furthermore, we should also be aware that due to price assessment, deviations from fundamentals may already be priced in *a priori.*

Mutual interdependencies in the global crude oil pricing system exist not only in the geographic but also in the time dimension. Figure 3.2 draws the spreads between the WTI one-month and four-month futures price, respectively, and the spot price in the past 15 years (futures–spot). They used to move within a +/–8 US dollar-range. These borders were broken around the price peak in 2008. Intuitively, the one-month spread is much less volatile than the four-month spread.

Daily data of futures and spot prices in this time span reveal a clear-cut relationship. The correlation of the one-month futures price and the spot price is 0.9999. The four-month futures price and spot price correlate at a value of 0.997. Thus, both values are hard to distinguish from 1. Since open interest in futures contracts is strongly concentrated within contract lengths of one to four months, we can say that futures and spot prices closely follow the same path. In accordance with Figure 2.1, arbitrage opportunities between futures and spot prices are exploited such that they are equal alongside the suggested structural differentials.

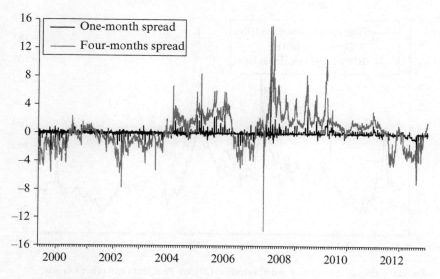

Source: Energy Information Administration (2015a). Petroleum and other Liquids.

Figure 3.2   *One-month and four-months WTI futures–spot spreads in US dollars, 2000–14*

3.3 THE RELATIONSHIP BETWEEN CRUDE OIL AND OTHER FOSSIL ENERGY SOURCES

Besides the geographical and temporal dimensions, the degree of global integration of the crude oil market is also determined by its relationships to other sources of energy. It is relevant to know if an event in the crude oil market is limited owing to its isolation, or if it affects the market for energy as a whole. Let us consider the other two fossil fuels, that is, coal and gas. Figure 3.3 shows their weekly prices in comparison with the crude oil price pattern. Price data of natural gas of the Henry Hub type is provided by the EIA (2015c). The global coal price index is calculated by the Hamburg Institute of International Economics (HWWI, 2015). The overall impression is that fossil fuel prices move more or less together. While all three series tend to move closely along with one another from 2000 until 2008, a discrepancy opens up between coal and oil on the one hand and natural gas on the other. A small set of numbers can give an impression of how the price series might be connected. The coefficient of correlation between crude oil and natural gas prices is 0.12 for the whole sample. This is positive but quite small. If we take only the period from 2000 until 2008 into

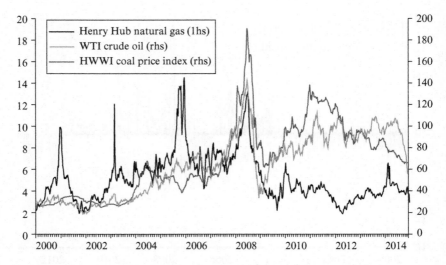

Sources: Energy Information Administration (2015a). Petroleum and other Liquids;
Energy Information Administration (2015c). Natural Gas; Hamburg Institute of
International Economics (2015). Coal Price Index, Datastream.

Figure 3.3   Natural gas, coal and crude oil prices, 2000–14, in US dollars

account, the coefficient considerably rises to 0.72. Correlation between
crude oil and coal is 0.87 for the whole sample.

These relatively strong correlation values open up space for a couple of
arguments about how the relevant prices may be related. The connection of
oil and gas prices is investigated in much more detail in the literature than
that of oil and coal prices. Villar and Joutz (2006, pp. 4–5) argue that, from
a demand perspective, a higher oil price leads to substitution of natural gas
for oil, thereby raising gas demand and consequently the gas price. On the
supply side, there are several counteracting effects. Depending on the specific source of a certain production plant, natural gas can be a co-product
of crude oil. A growing oil price induces higher oil production, which
thereby may raise the supply of associated natural gas and hence lowers
the gas price. On the other hand, a higher oil price caused by increasing
oil demand increases the need for production factors like labour or drilling
rigs. Higher prices of the latter raise production costs and lead to a higher
price of natural gas. Finally, a higher oil price has a positive effect on cash
flows of oil producers. They may invest it in new gas extraction projects,
raising gas supply and lowering the price of gas.

By investigating the time period from 1989 until 2005, Villar and Joutz
(2006) find a stable cointegrating relationship between the WTI crude oil

price and the Henry Hub natural gas price. Moreover, the oil price seems to be weakly exogenous, implying that oil influences gas rather than the other way around. However, this estimate does not cover the period when natural gas and crude oil prices start separating. Erdös (2012, p. 717) argues that a strongly rising supply of shale gas production took place from 2009 onwards, so that liquefying and export capacities got scarce. This prevented global arbitrage and gas price re-equalization and hence led to a drop in the Henry Hub price relative to the oil price. According to this idea, it is only a question of time until cointegration between oil and gas prices is restored, that is, a question of how long it lasts for transport capacities to adapt to production. The drop of oil and coal prices towards the end of 2014 may be a first sign of this tendency. Brigida (2014) allows for switching between different states in the relationship between crude oil and gas prices and thus takes the temporal deviation into account. The cointegrating relationship is strengthened.

Applying all these time series tools to the relationship between oil and coal would be beyond the capacity of this chapter. We rely on Figure 3.3 for the basic argument that crude oil and coal prices seem either to be causally connected or at least to be affected by a common third variable. This suggestion is confirmed by the rather high correlation coefficient. Yet, the case of coal should be considered more cautiously, since it is not as close a substitute for crude oil as natural gas.

It can be concluded that crude oil is not an isolated issue. Changes in the crude oil market are likely to have effects on prices and quantities of other energy sources. For example, a high oil price carries with it the probability of triggering the substitution of other energy sources for oil. Consequently, impacts of monetary policy on crude oil have further-reaching influences, notably on other energy markets. These effects might take place in the short or in the long run depending on the speed of market reactions.

3.4 THE DICHOTOMY BETWEEN US MONETARY POLICY AND THE GLOBAL CRUDE OIL MARKET

The US economy is the largest in the world. Its monetary policy has worldwide implications. Nevertheless, there exists a dichotomy between monetary policy and the market for crude oil. While the former takes place within the frame of individual countries, the latter is globalized as shown above. There is thus the basic question whether monetary policy actions of a single country can have a significant influence on a market that stretches over the whole planet.

The leading role of the US economy also makes its monetary policy a leading one. Changes in the US monetary policy stance are argued to transmit to the rest of the world. It is observed that monetary variables like interest rates and asset prices have become more correlated across emerging market economies in past years, specifically in the course of US quantitative easing (Mohanty, 2014, p. 3). US monetary policy seems to have international effects that give rise to the existence of certain transmission channels to other countries. In general, five main channels are identified (see, for instance, Caruana, 2013; Takáts & Vela, 2014). The first one is the exchange rate. It is the same mechanism that we already addressed in the previous chapter. Among the conventional transmission channels, it is the only one that has an immediate international impact, as it can only exist if there is more than one country. Expansive US monetary policy in general and quantitative easing in particular lead to a depreciation of the US dollar against other currencies, creating a knock-on effect on exports, output and other variables of these countries. Nominal and real exchange rate appreciations of emerging market economies have been modest during quantitative easing programmes in the United States (Mohanty, 2014, pp. 4–5). Some countries, especially several oil-exporting countries, have pegged their exchange rate to the US dollar. Hence, the transmission of US monetary policy effects to these currency areas is a logical consequence, even though the implications are not inevitably proportional to those in the United States.

Another channel is the setting of policy rates of interest. Takáts and Vela (2014, pp. 54–56) show that target rates of many emerging economies are considerably correlated with the US federal funds rate. This means that the central banks of these countries react to interest rate cuts in the United States by lowering their policy rates proportionately. Obviously, this transmission channel counteracts the exchange rate channel. By reducing their interest rates, emerging market economies prevent their currencies from appreciating. This is why Mohanty (2014, p. 5) observes that not only policy rates of interest of countries with pegged exchange rates, but also those with flexible exchange rate systems, follow US monetary policy conduct. This relativizes the notions of pegged and flexible exchange rates as such.

Long-run interest rates, reflected by long-term government bond yields, represent a further international transmission channel. The difference to the policy rate consists in the fact that correlation of long-term yields is not up to central bank reactions but takes place through financial markets: lower yields in the bond market of one country spill over to other countries (Takáts & Vela, 2014, pp. 57–58). Hence, the effects that US expansive monetary policy and especially quantitative easing have on US bonds, apply analogously to foreign bonds. Asset prices increase and long-run interest rates fall.

Cross-border bank lending constitutes a fourth transmission channel. It gives a picture of how monetary policy conditions set in the United States lead to the worldwide expansion of US dollar bank loans. A low interest rate not only gives an incentive to demand a credit denominated in US dollars in the United States but also does so abroad. There has been strong co-movement of foreign credit to emerging market economies since 2001. However, it has lost shares to domestic financing (Takáts & Vela, 2014, pp. 61–62).

The portfolio channel is similar to the bank lending channel. Corporations in emerging market economies have strongly raised the emission of securities at the cost of bank lending in past years (Mohanty, 2014, p. 7). Low interest rates in the United States promote carry trade: low-cost credit denominated in US dollars is invested abroad. This is also facilitated by corporate securities that are issued in US dollar denomination.

Importantly, the international transmission channels not only lead to a spillover of monetary conditions from one country to another but also spread financial risk. An expansive monetary policy stance in the United States is suggested to have a similar financial market effect in impacted foreign countries. The balance sheets of banks and corporations grow longer and so do risk exposures. Once US monetary policy becomes more restrictive, the transmission channels work in the opposite direction. A sudden outflow of capital from emerging market economies jeopardizes the stability of their financial systems. Currencies depreciate and inflation is on the way to increase. The countries have to raise policy rates, too (Caruana, 2013, p. 2).

The studies investigate a considerably large selection of countries. So we should expect that individual country characteristics impede a clear-cut result about international monetary policy transmission. Indeed, the above-mentioned significant results normally only apply to a fraction of the countries under investigation. Ramos-Francia and García-Verdú (2014) employ an impressive number of empirical tests and find mixed evidence about how international transmission has evolved in the course of the financial crisis that broke out in 2008.

Empirical results taken alone are not proof of the existence of transmission channels. While the international effect of US monetary policy is seen as a push factor that raises liquidity, there are also pull factors, reflecting the comparative performance of countries and their ability to attract foreign capital (Mohanty, 2014, p. 10). This shows again the demand-determined nature of money, illustrating that demand for credit in foreign countries is a precondition for capital to flow there. Interpretation of cross-country correlations of interest rates and capital flows thus becomes more difficult. The pull factors of capital flows may imply that countries are in a similar phase of the business cycle or face common economic growth prospects. This is a possible reason for the correlation of interest rates and capital flows. Conversely, the degree to which interest rates and capital flows differ across countries may reflect the differences in economic performance. These sources of variation in interest rate and capital flow variables are at least not directly linked to monetary policy. Adding the push factor, that is, monetary policy in the United States, gives us an indication but no definite evidence. Interest rate correlation and capital flows may be due to monetary policy or they may not. Thus, even though the empirical results allow for some suggestions about the international effect of US monetary policy, they should not be taken as absolute.

The evidence presented is about emerging market economies, including Asian, Latin American, Eastern European and a few other countries. While this selection covers many economically important countries, it leaves out advanced economies other than the United States. It is even argued that US monetary policy serves as an approximation for advanced economies as a whole (Takáts & Vela, 2014, p. 51). On the one hand, this is increasingly appropriate, because advanced economies now contribute less than half of global output (Adams-Kane & Lim, 2011, p. 2). On the other hand, as Borio et al. (2011, p. 45) show, foreign credit denominated in euros and yen has grown since 2000, too, especially in the years prior to 2008. But US dollar credit flows still are clearly much larger and their growth rate exceeds those denominated in euros and yen. Nevertheless, monetary policy in the euro area and Japan probably exerts analogous effects on other countries. This again makes room for many counteracting effects. Yet, monetary policy in advanced economies in recent years has in general been quite expansive and thus points basically in the same direction. The ambiguity of monetary policy transmission between advanced economies leads us to suggest that it is at least not entirely absent. Neely (2011) examines the effect of US unconventional monetary policy on several advanced economies, to wit, the United Kingdom, Germany and Canada. His results show that quantitative easing in the United States had no effect on these countries' short-run interest rates but a rather strong effect on their long-term rates.

Overall, these considerations about international transmission of US monetary policy give us many hints and indications but also leave us with several indeterminacies. We do not have explicit causalities, though it is likely and reasonable that monetary policy in the United States has effects beyond country borders. Being led by such reasoning, we may state that there is at least partial international transmission of monetary policy: on the one hand, if other countries do not follow US monetary policy, the US dollar exchange rate is likely to change so that there is at least one channel of international transmission. This is due to the fact that crude oil is traded in US dollars internationally. On the other hand, if countries seek to avoid currency appreciation against the US dollar, they implement their monetary policy in accordance with that of the United States. In one way or another, US monetary policy should have international effects. Having these international aspects in mind allows us to deal with the dichotomy between the global oil market and national monetary policy. International transmission probably cannot make this contrast disappear, rather, but mitigates it.

NOTE

1. The term 'unconventional' monetary policy is criticized by some authors outside of the neoclassical school since it does not affect the endogenous nature of money (see, for instance, Lavoie, 2014, pp. 226–229). However, since the literature largely referred to uses the term, we accept it for convenience.

4. Empirical evidence: monetary policy impacts on oil market variables

The aim of empirical analysis is to test the quality of theoretical ideas. It is useful to put abstract concepts into the specific historical context of real economies. However, it cannot be seen as an act of proving or disproving hypotheses with mathematical precision. Econometrics is rather a further argument in favour or against an economic theory. It does not describe physical laws but the real world where humans live. Humans are not mechanical but conscious beings. As such, they have individual behaviour that varies over time and depends on other individuals' behaviour (see, for instance, Bénicourt & Guerrien, 2008, p. 72). Estimating final economic outcomes can be done only with some – and usually considerable – degree of imprecision. The argument can be strong or weak depending on resulting evidence. On the one hand, it is hard to defend an economic idea if all existing empirical results deny it. On the other hand, we cannot be sure that a theory, which does not find support in real data, does not have any grain of truth. Likewise, seemingly convincing evidence does not guarantee the overall correctness of the underlying conception. There are many factors that complicate the finding of clear-cut estimation results or even make it impossible.

4.1 MEASUREMENT PROBLEMS: QUANTITATIVE ANALYSIS REQUIRES QUALITATIVE BACKGROUND

A first and omnipresent measurement problem in economics consists of time lags. Most if not all effects materialize with a certain delay in the rest of the economy. They impede empirical estimates, as we do not exactly know when a specific effect takes place and whether it occurs once in a point in time or only gradually.

A second difficulty in econometrics is the building of expectations by agents. This is what we say to give rise to uncertainty in economics. With regard to empirical analysis, it brings with it the problem that the impact of a variable change may not be found to be significant because individuals anticipate it. The effect then exists but it cannot be assessed from data. Monetary policy gives a famous example of this problem. Modelling a monetary shock that yields significant results is part of an ongoing debate. For instance, Bernanke and Mihov (1995) favour the federal funds rate of interest as a generally appropriate measure of monetary policy. Indeed, it has become the most conventional one. Another measure is proposed by Romer and Romer (2004), who try to filter out endogenous movements of the federal funds rate. The newly created data series is supposed to be relatively free of endogeneity and anticipation problems (ibid., p. 1056). Its drawback is that by extracting presumed endogenous components of the federal funds rate, a part of the monetary policy effect is extracted, too. For example, if investors anticipate a future change in monetary policy conduct and thus adapt their investment portfolio, the effect is forestalled. It is hard to detect it in data. But it is nevertheless an effect of monetary policy. The concerns about measuring conventional monetary policy apply equally to unconventional policy, as anticipation also occurs in view of asset purchases by the central bank.

Even further-reaching problems with respect to expectations arise in connection with supply, demand and price mechanisms. As we argued in the context of Figure 2.1, the curves can basically have any slope (see Pilkington, 2013). It thus becomes harder to find significant results as middle- or long-run developments to be tested are disturbed by erratic short-run noise in the data. An example of this is the liquidity preference of investors. We expect expansive monetary policy measures to stimulate economic activity and asset prices to some degree. Yet, as noted, the announcement of extraordinary monetary policy conduct may give agents the signal that economic performance is worse than they thought. They might react by a flight into money or liquid assets like government bonds. Prices of other assets fall therefore. Even if monetary policy is effective in reaching its goal of stimulating the economy, empirical data may reveal a counteracting short-run slack. Another example is the pattern of oil inventory accumulation that will be emphasized further in this chapter.

Furthermore, data on OTC trade is missing. Existing data inquiries are quite limited (Bank for International Settlements (BIS), 2013, pp. 9, 12). We may suggest that NYMEX Light Sweet Crude Oil Futures data are representative of the oil derivatives market, as global integration eliminates arbitrage opportunities. This is an appropriate working assumption. But we should be aware that OTC trade leaves potentially important information in the dark.

A specific characteristic of the crude oil market is its globalized competition on the one hand and its connection to political issues on the other. It is, in general, difficult to find stable relationships in complex economies.

In our particular case, it becomes even more difficult as the oil price may be directly affected by political decisions in OPEC countries. These decisions depend on specific strategies and historical circumstances. It is thus nearly impossible to model them. As a consequence, the oil supply curve gets a changed form. In particular, the price-quantity feedback mechanism may not be symmetric with regard to oil supply. Oil production of OPEC countries might react fundamentally differently to a price increase than to a price drop. For example, the OPEC countries may aim at competing against non-OPEC countries by influencing production quantities. This would imply a rise in production in order to drive private competitors to insolvency. Such may be the case in the second half of 2014, when OPEC countries did not decide to cut production even though the price was falling by about 50 per cent within a few months (see Krauss & Reed, 2015; Reed, 2014, 2015). Hence, for example, OPEC production might stay constant when the oil price is rising or it may rise when the oil price is falling. This reaction, or non-reaction, is not primarily due to low production flexibility but to political strategy. Such asymmetries complicate any empirical analysis, because they cover or partially replace the market mechanism, which is quite often suggested to follow a regular pattern.

A further, similar problem about the oil supply curve arises from the fact that crude oil is a natural resource. Oil reserves differ in quality and accessibility. The final commodity supplied has therefore been produced at different costs (IEA, 2008, pp. 217–219). As such, we cannot assume that all oil demand can be satisfied under equal conditions, which means that the long-run supply curve is not horizontal. The higher the demand, the more easily accessible oil reserves are exhausted and the more oil with higher production cost has to be explored. This means that the oil supply curve is probably not linear. Within a particular range of the oil price, oil supply may not react at all to a price increase, so that the curve is vertical in this range. In another range, where we assume that the price reaches a level where a production technology for a specific type of oil becomes profitable, a small price rise may trigger a quite large increase in production. Investment behaviour of oil companies therefore is probably non-linear, too.

To draw a conclusion that brings us a step further, we need assumptions and simplifications to overcome empirical measurement problems. These assumptions may hamper the finding of significant results. So, besides the advantages that empirical models bring, there are the disadvantages of missing details. It is only a theoretical analysis that is able to make sense of a complex issue without having to make strong assumptions. Thus, in the course of our empirical investigation, the theoretical analysis remains our benchmark.

4.2   AN SFC MODEL OF THE CRUDE OIL MARKET

Complexity of the real world requires simplification, if one wants to draw empirical conclusions. In this chapter, an SFC model of the crude oil market is presented in order to crystallize the main effects and principal variables. We follow the basic principles of modelling discussed by Godley and Lavoie (2012). The outstanding characteristic of these kinds of models is that they are truly macroeconomic in the sense that model results are not based on strong assumptions of individual behaviour that determine an aggregate equilibrium. In SFC models, deviations from stationary points merely induce reactions whereby the economy adapts to new conditions. This reflects the fact that market participants make their decisions in a world of uncertainty (Godley & Lavoie, 2012, p. 16). Hence, room is left open for a large set of possible outcomes. Moreover, SFC models are consistent with reality in that they take into account the double-entry characteristic of all economic stocks and flows. This allows for an appropriate monetary approach, because it is money that implies the relationship of all financial assets to an equal amount of debt. For example, by this proceeding, we respect the fact that financial investors do not just behave according to developments in fundamentals or asset prices. Their investment decisions are also affected by the cost of investment, that is, the interest rate on borrowed capital that is used to purchase assets.

The model at hand serves as a summary of the theoretical analysis and yields a rationale before we start with our empirical examination. The modelling of the oil market takes place at the cost of details that we have emphasized above, like, for example, those of the oil price assessment process. Nevertheless, it is able to reach an integration of the futures and spot market with dynamic interactions. As such, it serves as a tool to show basic mechanisms. Moreover, it gives us an impression of the results to which indeterminacy in capitalist economies can lead. We will see that the variation of model parameters can yield different results that hamper econometric testing. Therefore, the model shows us also which empirical approaches promise to be successful and which do not lead to meaningful results. Besides reliance on general principles of SFC modelling, this model, which contains monetary policy, and its relationship to an integrated crude oil market, is new and has, to our knowledge, no forerunner model in the literature.

As a first crucial feature of the model, money is endogenous. The central bank sets the interest rate while the quantity of money is determined by demand for credit. There is real as well as financial investment. Since the SFC model does not have a single and ever valid equilibrium as a gravitation centre and since decisions are made under uncertainty, the efficient markets hypothesis loses its reference point. As a second feature, therefore, the model gets along without the efficient markets hypothesis.

The model itself does not provide empirical results. The parameters are set according to economic reasoning concerning their sign but do not reflect the exact values achieved by empirical calibration. Hence, the model is constructed rather than estimated. It thus serves as a starting point for the econometric analysis and lines out the individual relationships that have to be estimated.

4.2.1 The Model Structure

The SFC model contains 43 endogenous variables and hence consists of 43 equations.[1] It starts with the equation of the crude oil spot market. Demand for oil is assumed to depend on two variables, that is, economic output and the oil price:

$$C_{oil,\, d} = \delta_0 + \delta_1 {}^* C_s - \delta_2 {}^* p_{spot} \tag{4.1}$$

where $C_{oil,d}$ is demand for oil, C_s is non-oil output and p_{spot} is the spot price. C_s is taken as exogenous. A variation in non-oil output thus represents a change in oil demand that is rooted in economic fundamentals. The equation shows how the oil market evolves if there were no futures market distorting fundamental developments. We implicitly assume that the demand curve is falling, that is, δ_2 is positive. This assumption may be reasonable both in the short and long run but may as well be violated temporarily. We will come back to this aspect.

Oil production is divided into oil sales and accumulation of inventories. This is a logical fact: oil production amounts to a given quantity. Oil supply, in contrast, is only what is effectively offered at a given market price. Therefore, it is equal to oil sales. The difference is the change in oil stocks:

$$\Delta IN = \delta_3 {}^* K_{-1} - \gamma {}^* IN_{-1} - C_{oil,s} \tag{4.2}$$

where IN is inventories. K is the capital stock in the oil industry at the end of the last period that has become operative in the current period. Consequently, δ_3 is a measure of production technology that creates a relationship between capital and crude oil produced. This simple modelling of oil production does not mean that we leave labour out of consideration. It will be shown in another equation that capital and labour input both grow proportionally to oil output. The concentration on capital in equation 4.2 facilitates the modelling of oil industry investment.

One should be cautious about the term $\delta_3 * K_{-1}$. It does not represent effective oil production but rather production capacities. The amount to which capacities are used is determined by the level of inventories of the previous period in equation 4.2. The higher the stock of inventories, the easier an oil company can react to unforeseen changes in oil demand. If existing oil stocks are already high, there is no need for a company to accumulate any further. Production capacities then do not have to be fully utilized. Production in excess of oil sales is hereby reduced. The coefficient γ thus determines the degree to which the inventory level translates into capacity utilization.

As discussed in abundance, supply and demand are always equal in any market. This means for the oil market and for the rest of the economy that

$$C_{oil,s} = C_{oil,d} \tag{4.3}$$

$$C_d = C_s \tag{4.4}$$

The equality between oil supply and oil demand always holds. This signifies, in connection with equations 4.2 and 4.1, that demand is the driving variable and determines capacity utilization and inventory accumulation.

The sum of non-oil output and oil production yields total output. The model is based on an unfamiliar notion of GDP, Y, which is taken in both nominal and real terms. In particular, it is real with respect to economic output other than crude oil but nominal with regard to crude oil production. It consists of

$$Y = C_s + p_{spot} * C_{oil,s} + I_s + p_{spot} * \Delta IN \tag{4.5}$$

Hence, exogenous non-oil output, C_s, and oil industry investment, I_s, are in real terms. Oil production, that is, effective oil supply and the change in inventories, are measured in terms of the current price. It follows that we are not interested in the absolute oil price level but in the oil price in proportion to the prices of the rest of the economy. In other words, we assume the prices of non-oil production and oil investment goods to be constant but allow for the oil price to vary. This is in line with the framework applied in the theoretical analysis of the monetary transmission channels in Chapter 2, as it gives rise to oil price changes relative to the general price level. Importantly, let us recall that we have a variable of total output, Y, but that the focus of the model is on the oil industry. The fact that non-oil output is exogenous means that all associated non-oil variables are exogenous, too. Indeed, we ignore not only non-oil prices but also non-oil investment, non-oil wages, non-oil capital, non-oil loans and so on. There is no logical inconsistency about this. In the following, all variables besides C_s concern the oil industry rather than the whole economy.

Investment demand in the oil industry depends on expected profits of future oil sales and risk exposure:

$$I_d = \frac{\alpha_1 * PP_P^e}{1 + \alpha_2 * L_{P,-1}} \tag{4.6}$$

The higher the expected profits, PP_P^e, the stronger is the incentive to enlarge production capacities, to wit, to raise investment spending. L_P is bank credit taken out to finance production; the larger the amount of credit, the higher the leverage and the higher thus the risk of going bankrupt. Producers hesitate to increase risk without limit. Additional investment, however, requires further credit and thus raises risk even more. Hence, high indebtedness constrains investment in order to prevent leverage to keep growing to infinity. Expected profits are simply given by profits of the preceding period:

$$PP_P^e = PP_{P,-1} \tag{4.7}$$

Production profits are obtained by simple subtraction of cost from return. A share s of them is distributed to the owners of oil-producing firms while the rest is reinvested:

$$PP_P = Y - C_d - W_d - r_{-1} * L_{P,-1} \tag{4.8}$$

$$PPU_P = (1 - s) * PP_P \tag{4.9}$$

where PPU_P represents undistributed profits. Even though the interest rate level is not an immediate determinant of investment in equation 4.6, it has nonetheless a strong influence as it directly impacts on profits in equation 4.8. Thereby, expected profits are equally affected, so that the interest rate finds its way to the investment equation. This is the effect of monetary policy through the fundamentals of the oil market. Investment demand is, naturally, equal to the supply of investment goods. Moreover, investment consists of the equipment that is added to the capital stock. We abstract from capital depreciation.

$$I_s = I_d \tag{4.10}$$

$$\Delta K = I_s \qquad (4.11)$$

Wage rates, denoting wages in proportion to total output, are assumed to be constant. Labour input, W_d and thus total wages paid increase in proportion with oil production. δ_7 in equation 4.12 summarizes the labour share of income and a technology measure transforming labour into oil produced. Demand for labour is again equal to supply of labour, W_s.

$$W_d = \delta_7{}^*(C_{oil,s} + \Delta IN) \qquad (4.12)$$

$$W_s = W_d \qquad (4.13)$$

Let us now turn to the oil futures market. While the spot market is more determined by middle- and long-run developments, the futures market is quite short-lived. Hence, as a fact to be well aware of, making assumptions about speculative behaviour is not an easy task. Futures contracts are traded by a long side and a short side. Short positions, F_S and long positions, F_L, are equal in every moment:

$$F_S = F_l \qquad (4.14)$$

We assume that producers exclusively go short while financial investors only go long. On the one hand, this assumption is strong, as we will emphasize later. On the other hand, real data show that net positions are as we argue (CFTC, 2014). If we cover all futures market traders other than producers by the term 'financial investors', we are not too far from reality. Even though this idea will not be that easy to obey in econometric tests owing to the composition of datasets, it is logically consistent with our theoretical analysis: what matters is demand pressure exerted by financial investors. Furthermore, we cross out offsetting positions that producers and investors hold among themselves.

Another equality has been derived in our theoretical analysis. We argued that oil prices in the spot market and in the futures market are equal. We abstract from 'contango' and 'backwardation' situations, since we do not assign them causal power in driving prices. Thus, developments in the spot market have a direct impact on the futures market and *vice versa*.

$$p_{spot} = p_{fut} \qquad (4.15)$$

where p_{fut} is the futures price of crude oil. It is driven by the amount of open interest, by developments in market fundamentals, and by the propensity of producers to rely on futures market demand:

$$p_{fut} = \frac{\delta_4 + \delta_5 * (F_L + C_{oil,d})}{\delta_6 * K_{-1}} \tag{4.16}$$

In equation 4.16 the term within brackets measures the total demand force that acts on the futures price. δ_5 thus is a parameter of price sensitivity. Intuitively, the futures price increases with demand power exerted in the futures market, F_L. The second term, spot demand $C_{oil,d}$, is less clear at first sight but also quite intuitive. We know, and argued in detail, that price-driving speculation in the futures market is likely to reduce spot demand for oil. This occurs since the futures price is equal to the spot price while the spot price exerts its effect on oil demand in equation 4.1. This has a price-lowering effect. The reduction in spot demand that we model by the variable $C_{oil,d}$ can in reality materialize in two ways. Either financial investors recognize it by observing the spot market such that they reduce their long position bids. Or consumers hold long positions with a hedging instead of a speculative intention. By reducing consumption demand, they reduce their futures position held. In both cases, demand increase by financial investors in the futures market has to exceed the decline in spot demand in order for the price to keep rising. The supply side has a decreasing effect on the price, too, because producers are willing to accept a lower price for futures contracts in the face of decreasing spot demand. By introducing $C_{oil,d}$, we model these effects in an indirect way. Equation 4.16 is indeed the key equation connecting the spot and futures markets. F_L is determined in the futures market but has an effect on the spot market through the futures price, which is equalized by the spot price. On the other side, $C_{oil,d}$ is assessed in the spot market but has an impact on the futures market, because its change promotes or hampers speculative behaviour. It is the double nature of oil as a commodity and a financial asset that closely links both markets.

Price changes are mitigated to a certain degree by installed production capacities, K_{-1}. Oil producers have to decide whether they accept the price bids of speculators in the futures market. Let us assume the case where oil is available in abundance and any further unit can be supplied with little additional effort. In this case, competition forces producers to accept relatively low bids. In the opposite case, where oil is scarce, the more that producers have difficulties in supplying additional units of oil, the higher the price that they require. The lower production capacities, the harder it is to satisfy any demand. This leads to a stronger futures price reaction in the face of increasing demand.

Financial investment in the futures market is highly leveraged. We argued that the trade of futures contracts creates high amounts of debt owing to the fact that the trader is currently in possession of an asset but the commitment to pay is given for a specific date in the future instead of the present. All one has to pay in advance is a maintenance margin. The leverage then consists of the debt – or, as we argued, a virtually created unofficial form of money – over the maintenance margin. The rates of margin requirements usually are between 10 per cent and 20 per cent of the contract value (see, for example, Investopedia, 2015). The maintenance margin to be held for trading a given number of futures contracts therefore depends on the price level in the futures market.

$$F_L = \frac{M_I}{m^* p_{fut,-1}} \tag{4.17}$$

where M_I is the amount of the maintenance margin held by financial investors. m is the rate of margin requirement. The higher the futures price, the less futures contracts an investor can buy with a given amount of margin. Alongside this simplification, equation 4.17 is a logical accounting equality and does not require further modelling assumptions. However, we leave away the maintenance margin of producers when they go short in the futures market in order to hedge their sales. The loss of model quality is limited. Equation 4.17 would have exactly the same form for producers apart from F_L becoming F_S. One may argue that producers, that is, the short side in the futures market, only have a passive role and automatically accommodate financial investors' demand for futures owing to equation 4.14. Thus, they would participate in the futures market according to investors' wishes irrespective of the potential losses that futures trading may bring them. This is not true. The willingness of producers to agree on futures contracts depends on the futures price that investors are willing to pay. The conditions of producers' futures supply are expressed in equation 4.16 by parameter δ_5 and K_{-1}.

Every purchase of a futures contract requires a corresponding margin balance. Consequently, continuing from equation 4.16, financial investment in the futures market is expressed by an increase of the maintenance margin. Financial investment depends on expected profits, the interest rate and risk exposure:

$$\Delta M_I = \frac{\beta_0 + \beta_1^* FP_I^e - \beta_2^* r}{1 + \beta_3^* L_I^2} \tag{4.18}$$

where FP_I^e is expected financial profits of investors and r is the interest rate. The effect of the interest rate has two complementing interpretations. First, a lower interest rate makes borrowing cheaper. Investment in futures contracts can be made at a lower cost. Second, the interest rate affects liquidity preference. L_I is bank credit for financial investors. The larger the indebtedness of investors, the higher the risk exposure. The argument fully corresponds to failure risk of oil producers in equation 4.6: given a high risk exposure, investors become more cautious of taking additional credit. Raising the weight of risk by squaring the variable is an issue of modelling but has a reasonable explanation. Real investment in the oil sector creates a real asset that backs investment expenditures. In contrast, financial investment is only backed by paper oil, which consists in large parts of an uncovered leverage. Loans granted to financial investors are therefore more risky than those granted to producers. This is why we square investors' debt but do not do so with producers' debt. All in all, equation 4.18 replicates the financial market effect of monetary policy on the oil market.

The amount of credit that speculators have to borrow is given by the desired capital to invest and the portion of it that can be contributed out of their own means:

$$\Delta L_1 = \Delta M_1 - FPU_1 \tag{4.19}$$

FPU_I is the realized and undistributed financial profits of investors that are available for reinvestment. The need for credit is reduced by the same amount. Total financial profits are calculated by changes in the level of long positions and changes of the spot price in comparison with the agreed futures price. The borrowing cost is subtracted.

$$FP_I = (p_{spot} - p_{fut,-1})^*\Delta F_L + \Delta(p_{spot} - p_{fut,-1})^*F_{L,-1} - \frac{r_{-1}}{52}^*L_{I,-1} \tag{4.20}$$

The first two summands in equation 4.20 reflect how changes in the number of futures contracts and prices yield financial profits. Godley and Lavoie give the formula and the proof of why they represent all sources of asset returns (2012, pp. 134–136). If the spot price has grown higher than the futures price of the past period when the contract was made, financial investors make a profit if given that new contracts have been made, that is, ΔF_L is positive. The second summand in equation 4.20 represents how the difference between the futures price and the spot price has changed for positions that existed already in the last period or have been rolled over. The last summand in equation 4.20 is the interest cost for the credit that was necessary to finance financial investment. The interest rate is divided by 52, because we run the model in weekly frequency and rely on the convention that interest rates imply a payment once a year.

Expected profits are defined analogously:

$$FP_I^e = (p_{I,spot}^e - p_{fut,-1})*\Delta F_{L,-1} + \Delta(p_{I,spot}^e - p_{fut,-1})*F_{L,-1} - \frac{r_{-1}}{52}*L_{I,-1}$$
$$(4.21)$$

where $p_{I,spot}^e$ is the spot price expected by financial investors. Expectations are built at the beginning of the period. We thus assume that investors base them on the number of futures contracts of the past period. The expected spot price relies also on past price developments:

$$p_{I,spot}^e = p_{spot,-1} + (p_{spot,-1} - p_{spot,-2})$$
$$(4.22)$$

This means that investors take the spot price of the previous period as the initial point but build as well an expectation of the direction in which it may move. The expected price change is given by the actual change that occurred between the two preceding periods.

We assume that financial profits are fully reinvested, that is, undistributed profits are equal to total financial profits. This may appear as a strong assumption. But if a positive amount of financial profits were redistributed, this would hamper financial investment only to the extent that it reduces expected financial profits. Financial investment is basically determined by equation 4.17. However, we note for completeness that

$$FPU_I = FP_I$$
$$(4.23)$$

where FPU_I represents undistributed profits of financial investors.

Financial profits of oil producers are calculated the other way around. They benefit if the spot price has fallen below the level of the agreed futures price one period before. In analogy to investors' financial profits, we assume that producers fully reinvest financial profits.

$$FP_P = (p_{fut,-1} - p_{spot})*\Delta F_s + \Delta(p_{fut,-1} - p_{spot})*F_{s,-1} + CG_P$$
$$(4.24)$$

$$FPU_P = FP_P$$
$$(4.25)$$

We leave away credit cost in equation 4.24 since we ignore the maintenance margin of producers above. To remind, their credit needed for production is taken into account in the profit equation 4.8. In contrast, producers' capital gain, CG_P, is introduced. It accrues from a change in the valuation of oil inventories:

$$CG_P = \Delta p_{spot}*IN_{-1}$$
$$(4.26)$$

After this, the producers' need for credit can be determined. It is the sum of what is needed for production and investment minus their own financial means out of undistributed profits.

$$\Delta L_P = I_d + \Delta W_s - PPU_P - FPU_P + CG_P + p_{spot}*\Delta IN \qquad (4.27)$$

Oil production requires credit to finance wages. The change of credit therefore changes to the extent that labour employed changes compared to the previous period. Production and investment spending can partially be financed out of production and financial profits. Capital gains enter equation 4.27 positively, because they are contained in FPU_P but do not effectively contribute to liquid cash for expenditures. Capital gains rather appear in the illiquid form of inventories. The same applies to the changes in the level of inventories.

Let us turn to the banking system. Total loans that banks grant to customers are determined by the sum of loans to oil producers and loans to investors:

$$L_B = L_P + L_I \qquad (4.28)$$

where L_B is the total volume of loans granted by banks. They yield a rate of return equal to the market rate of interest. Bank profits are therefore given by the difference between the interest payments received from borrowers and interest payments that banks have to pay on central bank reserves:

$$P_B = r_{-1}*L_{B,-1} - r_{T,-1}*R_{B,-1} \qquad (4.29)$$

where $R_{B,-1}$ represents central bank reserves, which are held on the central bank account. r_T is the interest rate on them. We assume that the futures market clearing house is part of the banking system, so that the maintenance margin of financial investors is a kind of deposit that does not bear interest earnings. Credit taken for oil production is suggested to circulate in the form of cash in the hand of workers and consumers as will be seen in a moment. The banks' need for reserves is thus given by

$$\Delta R_B = \Delta L_B - \Delta M_B \qquad (4.30)$$

M_B is the maintenance margin of investors that takes the form of a kind of deposit in the view of banks. Hence,

$$M_B = M_I \qquad (4.31)$$

Monetary policy is made by targeting r_T, the interest rate of the interbank market. Central bank profits are simply given by the return on reserves paid by banks.

$$P_{CB} = r_{T, -1} * R_{B,-1} \qquad (4.32)$$

As equation 4.29 shows, banks' profitability is traditionally determined between the interest rates on their loans and the interest rates they have to pay on deposits and central bank reserves. Equation 4.33 thus defines an interest differential that we argue to be realized in the market:

$$r = r_T + D \qquad (4.33)$$

We assume that the interest differential D is exogenous. This is realistic in the sense that as long as the transmission mechanism of monetary policy works, the difference from the interbank rate to market rates of interest should be more or less constant in order to guarantee profits.

Consumers are the remaining group in the model. They generate income by working in the oil industry and by receiving profits distributed by producers, banks and the central bank. Both banks and the central bank distribute their profits completely to the public, so that they end up in the pocket of individuals who also have the role of consumers. We call the sum of these profits the 'profits of consumers':

$$P_C = s*PP_P + P_B + P_{CB} \qquad (4.34)$$

Consumers hold cash at the end of the period to the extent that income exceeds expenditures for oil consumption:

$$\Delta H_C = W_S + P_C - p_{spot} * C_{oil,d} \qquad (4.35)$$

where H_C is cash held by consumer households.

For formal completeness, two accounting equations have to be added. First, cash created by the central bank is equal to cash held by consumer households, since there is no cash held by another entity. Analogously, the reserves provided by the central bank are equal to the reserves held by commercial banks.

$$H_{CB} = H_C \qquad (4.36)$$

$$R_{CB} = R_B \qquad (4.37)$$

Finally, producers, financial investors, banks, the central bank and consumers have a certain amount of wealth at the end of every period. Producers' wealth consists of the capital stock and inventories net of bank credit. Financial investors' wealth is given by the volume of capital invested, that is, M_I, also reduced by debt. The wealth of banks is zero. This is intuitive, because central bank reserves must be of an amount such as to fill any gap between loans and deposits, as shown in equation 4.30. Moreover, we assume that they distribute all profits so that no past profits contribute to the growth of equity. The wealth of the central bank is equal to reserves to banks minus cash held by consumers. Cash is central bank money that represents a fraction of reserves. Since this fraction of central bank money is therefore not in the hands of the monetary authority, it has to be subtracted from the total of reserves. Consumers' end-of-period wealth is the cash they hold. Summing up the wealth of all groups yields total wealth in the oil industry, which obviously consists only of real wealth, rather than both real and financial wealth. The only nominal element is the oil price, because of the notion of GDP adopted in our model.

$$V_P = K + p_{spot}{}^*IN - L_P \qquad (4.38)$$

$$V_I = M_I - L_I \qquad (4.39)$$

$$V_B = L_B - M_B - R_B \qquad (4.40)$$

$$V_{CB} = R_B - H_C \qquad (4.41)$$

$$V_C = H_C \qquad (4.42)$$

$$V = V_C + V_P + V_I + V_B + V_{CB} = K + p_{spot}{}^*IN \qquad (4.43)$$

To point out again: this model is of a closed character. Despite its considerable volume, it remains a theoretical model and is not able to incorporate complexities of the oil market sufficiently. Its strength lies in the ability to show basic mechanisms, to visualize key effects that have been elaborated upon in the preceding analysis. The model lays the basis for econometric analysis. Moreover, its weaknesses exhibit the actual difficulties in handling with complexity.

For example, the model suffers a certain time inconsistency. We divide the interest rate by 52 in equations 4.20 and 4.21 through the argument that we run the model in weekly frequency. There would also be arguments to adopt daily frequency, owing to fast moving events in financial markets. In other equations (equations 4.8, 4.29, and 4.32) concerning oil production, bank lending and monetary policy, we leave the interest rates unchanged. This is appropriate from the point of view that production takes place over a longer time horizon. Moreover, financial markets react, in contrast to fundamentals, to the smallest short-run price movements. Similarly, banking that only consists of granting loans is not that exciting, as developments would have to be documented weekly. The same applies to monetary policy. Yet, here a problem arises, because yearly interest rates are nonetheless applied to a model run at weekly frequency, which is a contradiction. By trying to integrate fast-evolving financial markets with longer-lagged production we have to make this concession. It does not make a crucial difference to model results, because the patterns of variables differ only in the lengths of the time lag according to whether interest rates are divided or not.

Two limitations concern the financial sector. Financial investors only have crude oil futures (net) long positions as an investment opportunity. Once the price falls, they inevitably end up with losses and uncovered debt, because they cannot rearrange their portfolio apart from varying the total volume of invested capital. However, we are not interested in profitability of speculation but rather in the effects speculation has on the oil market. Second, the model has an exclusive focus on conventional monetary policy. Nevertheless, we shall overcome this shortcoming as we argue unconventional monetary policy to be analogous in its effect on the oil market.

As another potential shortcoming, the character of crude oil as an exhaustible natural resource is neglected. Indeed, oil production behaves like the supply side of any goods market, reacting accommodatingly to changes in demand. The long-run supply curve is considered as horizontal. Criticism in this respect is justified. But as will be seen in the empirical part, accommodative oil supply is not an assumption too far from reality. The model thus can continue doing its work.

The effect of economic growth is described separately. For instance, it is relevant in the case of the financial crisis, involving a drop in global output and hence in the oil price. It is shown below that the model can deal with (positive and negative) growth rates.

As regards monetary policy, the exchange rate is usually taken into account. In our model, however, it is not considered, since we consider the crude oil market in its global extension where exchange rates of single countries in general do not play a great role. We argued with respect to the exchange rate channel of monetary transmission that besides an induced change in the oil price denominated in US dollars, there is no significant change in global crude oil production and consumption. Changes in exchange rates produce effects in a national economy that are counteracted by the rest of the world, whose currencies *ceteris paribus* face the exchange rate change from the opposite perspective. Once the dichotomy between national monetary policy and the global oil market is analyzed – as we have done in Chapter 3 – neglecting the exchange rate in our SFC model seems appropriate.

Since the model is of a macroeconomic nature and describes a particular sector of a capitalist economy, strategic behaviour and geopolitical strategies like those of OPEC are ruled out. For example, the falling oil price in the second half of 2014 and the fact that OPEC decided to keep oil production at a high level to keep market share (Krauss & Reed, 2015; Reed, 2014, 2015) seems to lie out of the reach of monetary policy. This may be true, but one may as well argue that the need for raising OPEC production only emerged as a reaction after increased global oil production rendered an already falling price. If so, then the essential mechanism is captured by the model. It is strengthened by the OPEC strategy. In addition, strategic behaviour is by definition irregular and thus hard to reproduce by a macroeconomic model. To reproduce basic effects of monetary policy on the crude oil market, we find justification in leaving OPEC aside.

4.2.2   Running the Model

Despite the rejection of the general equilibrium approach, calibrating a model requires the assumption of a stationary starting point. This is necessary to detect the isolated effects on which our investigation focuses. A stationary starting point inevitably has the characteristics of equilibrium. However, once the starting point is passed, we are interested in the direction, length and strength of occurring effects rather than in perfect convergence to a new equilibrium. As we do not attest our model to be able to reproduce real data outcomes, it is calibrated with arbitrarily chosen values of variables. The signs of parameters are chosen as intended in each equation by economic theory but their numerical values are constructed rather than estimated empirically. However, the exact numbers of results are not of interest. What we want to reveal is the actual existence of effects.

The model is run from 1990 until 2014. The time frame is arbitrarily chosen. If we decided to run it at a daily instead of a weekly frequency, the same number of periods would be concentrated within about two or three years. In the equilibrium at the start of the time period, there is no futures market investment, producers' profits, real investment and inventories are zero. This may appear too artificial. However, this state may also be transferred to the real world in the sense that producers realize a minimum profit rate that is just sufficient to keep up production with neither investment nor disinvestment. Likewise, there may be inventories that oil companies hold in average to hedge against future risks. They are only carried over from the past and thus do not imply any price effects. Setting these values to zero at the beginning implies that our interest is in the deviation of the variables from a given initial point.

In the next step, a cut in the interest rate target from 3 per cent to 2 per cent at the beginning of 1991 is simulated. Figure 4.1 shows the evolution of important model variables arranged in an order in which variables cause other variables. Open interest, to wit, total long or short positions, is presented in panel (a). When the interest rate falls, investors start investing due to lower liquidity preference and reduced opportunity cost of futures investment. Note that both effects are at least partially of a permanent character since the interest rate remains permanently at the new lower level, too. Panel (b) exhibits oil industry investment sharply increasing after the cut in the interest rate. It soon reaches its peak and falls a little below zero, so that it is hard to recognize it in panel (b), in order to converge to the initial level of zero investment.

It is important to know that open interest in panel (a) shows the total volume of financial investment in the form of oil futures. Panel (b), in contrast, shows the amount of investment expenditures being spent again and again every period. Hence, open interest is a stock of capital while oil industry investment is a flow of capital. This means that total capital does not fall back to its initial level when investment reverts to zero. Rather, the picture shows that the capital stock of oil producers first grows strongly, then declines a little as long as investment is below zero, and eventually converges to a value above the initial one. As already mentioned in the context of equation 4.8, a lower interest rate raises expected profits and thus makes investment profitable. According to equation 4.6, the capital stock increases until expected profits again reach zero, owing to large and perhaps even excess production capacities.

Panels (a) and (b) represent the two broad ways of monetary policy transmission: the financial market effect and the effect through fundamentals, respectively. Yet, they are not the isolated effects directly caused by monetary policy. They include also mutual impacts on each other.

Panel (c) in Figure 4.1 gives a picture of the oil spot price evolution. Increasing demand in the futures market due to investment in futures contracts raises the price sharply as shown in equation 4.16. From equation 4.1, we know that spot demand decreases when the oil price increases. Declining spot demand affects in turn the futures price in equation 4.16 owing to financial investors becoming more cautious and producers being ready to accept a lower price in futures contract agreements. Moreover, and crucially, investment takes place and this increases production capacities. The supply side of the oil market becomes more and more relaxed such that oil companies are willing to drive up oil sales even at lower prices.

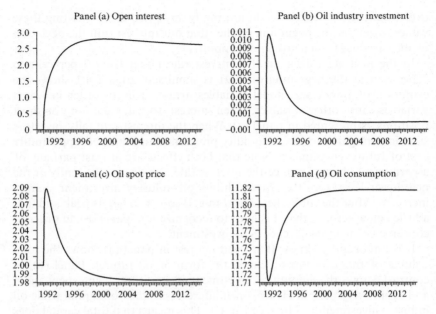

Source: Author's elaboration.

Figure 4.1   Effect of a 1 per cent decrease in interest rates

This is expressed in the denominator of equation 4.16. All these factors pull down the oil price from its peak and let it converge to a new stationary level even below the initial one. This can simply be explained by the fact that the oil industry ends up with higher production capacities that have been installed during the high price period. They do not vanish – or merely to a small extent – after the price has come down. Note that there is a kind of slight overreaction: the price falls deeply and then converges very slowly to the new level from below.

In panel (d), oil consumption is plotted. Consumption is what is effectively purchased in the market and not what is produced, and is thus the value of supply and demand of oil in every period. The higher oil price lowers oil demand, while price reversion pulls it up to a quantity larger than at the beginning. Since output of the rest of the economy is exogenous and remains constant during the whole time span considered, oil intensity of GDP depends uniquely on the absolute volume of crude oil consumption. In this case, the economy ends up with a higher oil intensity.

The effect of monetary policy on total GDP is neglected in this simulation, because we assume output of the rest of the economy to be

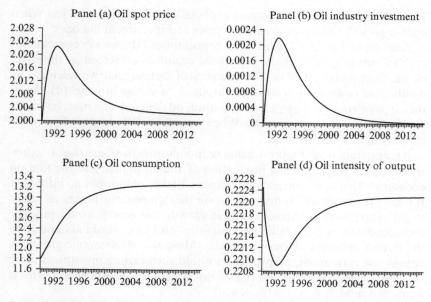

Panel (a) Oil spot price

Panel (b) Oil industry investment

Panel (c) Oil consumption

Panel (d) Oil intensity of output

Source: Author's elaboration.

Figure 4.2   Effect of economic growth

exogenous. We do so because we are interested in the effects of monetary policy on the oil market beyond its impact on the economy as a whole. Arguing that expansive monetary policy has a larger or smaller positive effect on output taking a specific time span to materialize, the oil market is affected by increasing demand. We reproduce this separate effect by assuming that the economy grows by 0.0004 per cent in a week, which amounts to about 2 per cent in a year. This happens for two years, that is, from 1990 to the end of 1991. After that period, the economic growth rate declines gradually and converges to zero in 2014. The growing variable is non-oil production C_s. Monetary policy conduct can remain unchanged for this. Figure 4.2 shows the evolution of the oil price in panel (a). As long as annual GDP growth is 2 per cent, the oil price increases. This is due to rising oil demand in equation 4.1, which transmits to equation 4.16. Oil industry investment, presented in panel (b), rises in light of a growing oil price. Since oil producers base investment decisions on profit expectations, which depend on realized profits in the preceding period, oil investment is lagged (equations 4.6 and 4.7). Once the economic growth rate diminishes, investment is again lagged and thus in excess of output growth in a given period. The price thus starts decreasing and so does lagged investment.

This corresponds well to common explanations of business cycles. When output growth converges to zero, oil price and investment fall back.

Panel (c) of Figure 4.2 exhibits oil consumption. It grows in correspondence to output growth of the rest of the economy, corrected by the effect of the fluctuating oil price. Oil intensity of output that we calculate by dividing oil consumption by total output, Y, is shown in panel (d). When the oil price rises owing to growing output, oil demand also rises, but more slowly. The oil intensity shrinks. When output growth drops, oil intensity rises again to the initial level.

It is apparent from Figure 4.2 that economic growth affects the oil industry. But with respect to the proportion of the oil industry to the non-oil economy – that is, oil intensity of output – higher output has no influence. It is only the process of economic growth that temporarily lowers oil intensity. If there were permanent output growth, the model would produce an outcome of steadily falling oil intensity. This case would assume away the output volatility of business cycles. Moreover, if economic growth is considered permanent, oil producers would start raising investment more strongly, because the fear and risk of investment loss would be less. This is a case not modelled in this framework.

Figures 4.1 and 4.2 separate monetary policy effects. The first simulation merely shows the effect of monetary policy on the market for crude oil in proportion to the rest of the economy. This is the central issue of the whole analysis. The second simulation and the subsequent extension reflect the impact of absolute output levels on the oil market. There are various dynamic effects between the oil industry and the rest of the economy but they are not permanent. The figures all in all yield an argument that isolating the effects of monetary policy on the crude oil market from general monetary policy effects on the economy as a whole is an appropriate way to proceed. Hence, the results of Figure 4.1 continue to be at the centre of our interest. Yet, short-run fluctuations should not be forgotten with regard to neither general nor oil market specific effects. They make up for an important part of economic phenomena.

Model results presented here are quite intuitive. Even though calibration of starting values is an arbitrary issue, the model is able to reproduce what has been concluded in the theoretical investigation. First and crucially, financial markets do not only affect prices temporarily, they also influence them permanently. As a consequence, not only price variables are impacted but also quantities. This reveals the topic of our starting point about the nature of money and financial markets. Fundamentals and money should not be seen as two separated and parallel objects. The model confirms the suggestion that economic analysis should be made from a monetary perspective. It is through money that fundamentals and financial markets interact dynamically. The transmission mechanism of monetary policy has both an effect through fundamentals and a financial market effect whose interrelations can lead to unforeseen outcomes.

4.3   PROCEEDING IN TWO STAGES

Our theoretical analysis and the SFC model reveal many steps that take place when monetary policy affects the market for crude oil. The initial point is a change in the short-run interest rate by the central bank. We argue that monetary policy then affects the oil market through fundamentals, on the one hand, and through financial markets, on the other, producing price and quantity effects.

We know, of course, that all these effects take place simultaneously at a given point in time. They occur at different frequencies and with different time delays. Often, due to complex interactions, they cannot be separated from each other. This calls for integrated measurement methods.

For these reasons, the following procedural steps are proposed. In the first stage, we estimate the effect of monetary policy on the futures market and on spot oil supply and demand as well as the impact they have on the oil price. By doing this, the fundamental and the financial market effects of monetary policy are covered as well as the interactions between them. Yet, the focus lies on the price variable, because the price effect, especially through financial markets, materializes faster than quantity effects. Reasonably and as argued above, quantity effects take longer to realize than price effects. Hence, they should be emphasized separately. In the second stage, therefore, we investigate the longer-run impact of the oil price on quantities, that is, oil production and oil consumption. A central variable of interest will be how a change in the oil price influences real oil industry investment.

Our two-stage procedure is therefore as follows: the first stage investigates the transmission of monetary policy from the interest rate to the oil price, while the second stage examines the impact that the oil price itself exerts on oil quantities. Despite their probable simultaneity, we analyze the two stages separately, because they differ in lag length and data frequency. By this method, we aim to empirically test the results of the SFC model shown in Figure 4.1.

There are several concepts of how a causal relationship from one variable to another may be estimated (see, for example, Hoover, 2008). The best-known and most applied one is the idea of Granger causality. It claims that if a change in variable a occurs before a change in variable b, and if variable a bears information that is not contained in any other variable, then variable *a* has some causal effect on variable *b* (Granger, 2004, p. 425). Yet, causality can be suggested but there is no obvious way of how it can be proven. This is why this concept is called 'Granger causality' and not just 'causality'. We will refer to Granger causality, too, when analyzing relationships between data. But it is necessary to be cautious: while data may reveal a relationship saying that one variable is able to forecast another one, it is not possible to find out from available data why the information that a variable yields should be unique. It is at this point that we have to rely on economic theory.

Hence, there are three tools helping to confirm or reject suggestions made in this book: the more detailed suggestive results of the theoretical analysis, the hints of the SFC model, and econometric methods.

4.4 FROM MONETARY POLICY TO OIL PRICE[2]

This section concerns the first stage of our examination. By investigating almost exclusively the effects of monetary policy on the oil price, we largely ignore the impact on oil quantities for the moment and refer to it later. First, monetary policy and the oil market are simultaneously treated in VARs. The results will not be sufficient, so a more detailed analysis of causal relationships will be required. The outcomes, whether significant or insignificant, then can be used for further, more sophisticated investigations of causalities. Finally, the oil inventory argument is taken into account in order to test the phenomenon of speculation. We criticized those numerous explanations that use oil stock data to assess whether price effects of speculation exist or not. By taking this criticism as a base, we employ a new approach to inventories, showing how they can be productively used in order to have some explanatory power.

4.4.1 Some Basic Estimations

Monetary policy in the time period of investigation, that is, from 2000 until 2014, is marked by the already discussed change from conventional to unconventional monetary policy. The latter started taking place at the beginning of 2009, when the federal funds rate reached zero in the United States. The US Federal Reserve had already begun taking action by purchasing financial assets in large quantities. Therefore, the period is separated into two at the end of 2008. The first one is referred to as the period of conventional monetary policy, the second one as the period of unconventional monetary policy. We start by estimating a structural VAR (SVAR) for each period (see, for instance, Enders, 2014, pp. 313–317).

This approach is in general judged to be convincing, as it allows detecting various causal relationships and estimating counteracting effects among many variables. It is thus just a common method to apply to issues like the present one. However, we will see that this approach is not without criticism.

VARs face the same trade-off of any regression analysis in a stronger form: more variables and lags are potentially able to include more effects but reduce the quality of the model at the same time. Hence, a selection of variables is necessary and has to be founded on economic criteria. The financial market effect is suggested to materialize faster than changes in oil supply and demand. To cover them, the SVAR is run with weekly data.

The first period, from 2000 until 2008
In the SVAR for the period of conventional monetary policy, we choose the federal funds rate as the policy variable provided by the Federal Reserve System (2015). It is not perfect, as argued above. On the one hand, it can be anticipated. On the other hand, the impact of a change in the rate of interest is likely to occur only gradually over time. For example, financial investors may raise their exposure only if interest rates are not just low but have been low for a sufficient period, so that they have confidence that the level will continue to be low. In the case of an increasing interest rate, refinancing costs of financial investors rise only step by step the more that credits have to be renewed. As a contrary argument, the federal funds rate is the one and crucial variable by which monetary policy influences monetary conditions. Investors rely on this rate when making decisions rather than imagining some artificial constructed variable. Hence, despite these limitations, we try to generate significant estimates by including lags.

The second variable is the WTI crude oil spot price from EIA (2015c). One may deflate it by the US consumer price index to have the real price of oil. This is what theoretical considerations – investigating the oil market in proportion to the rest of the economy – would suggest. But, first, data on inflation rates are only available at monthly frequencies. Interpolating them would bring no additional information. Second, US inflation has not exhibited specific features since 2000 despite the fall in the price level after the outbreak of the financial crisis. The correlation between the nominal oil price and the real oil price deflated by the US consumer price index (OECD, 2015) is 0.993. So we cannot make great estimating errors in applying nominal instead of real data. It is intuitive that weekly changes in the oil price are hardly due to a change in the general price level. Third, we argue that in a monetary economy, the nominal and the real sides of the economy are integrated and cannot be split up (Cencini, 2003a, pp. 303–304).

To include the financial market aspect, a variable of speculative futures trading is added as a third variable. In particular, we use non-commercial net position data surveyed by the CFTC (2014). The CFTC (2015) defines commercial traders as those 'involved in the production, processing, or merchandising' of crude oil. Another criterion to qualify as a commercial is the incentive to hedge by holding a futures contract (CFTC, 2014). Non-commercials make up for the residuum of trading activity and consequently do not have contact with, and probably no interest in, physical quantities. They do not aim at hedging and hence can be considered as speculators. Yet, such a classification is not easy to make and has also been criticized. Masters (2008, pp. 7–8) complains that so-called index speculators enter the futures market by contracting commodity swaps, which are classified by the CFTC as commercials rather than non-commercials. The non-commercial category is therefore more a guideline than a clear-cut indicator. Their net positions, that is, non-commercial long positions minus short positions, are a variable that is often applied in the literature (see Alquist & Gervais, 2011; Büyükşahin & Harris, 2011). As an approximation, the larger that net positions are, the stronger is the speculative activity. A case of negative net positions then indicates a situation where financial investors in total effectively go net short, because they expect the oil price to fall further owing to changing fundamentals. The path of this variable is drawn in panel (a) of Figure 4.3. While non-commercials tend to have positive net positions, the difference between long and short only takes off after 2008. Negative values may also be due to the imprecise definition of the category of non-commercials. Panel (b) shows the evolution of total open interest on a clear move upwards.

There is another inconvenience about net positions being an indicator

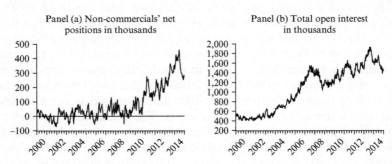

Source: CFTC (2014). Commitment of Traders: Historical Compressed.

Figure 4.3   Non-commercials' net positions and total open interest, 2000–14

of speculation. High demand of financial investors on the long side may be offset by short positions of other non-commercials. Even though this does not raise net positions, it may nevertheless have an effect on the price, because it is a sign of strong demand pressure to which futures supply adapts by way of an increasing price. Despite imperfections, we employ net position data to get an impression of a change in financial investment in the futures market that may be caused by monetary policy. Later on, this measure will be extended to include more potential indicators of speculative activity. We will see that it is insufficient to cover potential influences of financial investment. Other approaches to deal with speculation will be required.

Fourth, the fundamental side of the oil market should be included as well. The supply side of the oil spot market can be modelled by global oil production provided by the EIA (2015b). Herein, we follow a common path of the literature (see, for instance, Kilian, 2009b, p. 1058; Lombardi & Van Robays, 2011, p. 16). Yet, we will see later that this variable is not free of any problems. For oil demand, global industrial production data from the World Bank (2015) are chosen. Fossil energy being an important input, we suggest that an increase in industrial production raises the demand for crude oil. There are also other studies where this variable is applied in this sense (see, for example, Lombardi & Van Robays, 2011). Unfortunately, these variables are only available at monthly frequency. They are interpolated linearly at the expense of the information content of the series.

To keep the model parsimonious, the supply and demand sides of the oil spot market are merged into the ratio $\frac{industrial\ production}{oil\ production}$. If oil demand rises compared to oil supply, this ratio tends to increase. This, by the way, reflects changes on supply and demand sides and does not distort the discussed necessary equality of oil supply and demand themselves. If the oil industry reacts more (or less) strongly to monetary policy than the rest of the economy, the denominator should change more (or less) than the numerator. The fundamentals ratio should thereby be capable of taking the spot market into account.

The final variable is the exchange rate taken from the Board of Governors of the Federal Reserve System (2015d). It is an index of the US dollar against a broad group of important US trading partners. It may be surprising to learn why the exchange rate as a particular transmission channel is modelled with a single variable while other channels are not included explicitly. The exchange rate channel has – in contrast to the other channels – a quite direct link to the crude oil price, since the latter is traded in US dollars. As such, many other effects express themselves in a change of the exchange rate, which is expected to find its way to the oil price. This variable choice, of course, does not mean that all other transmission channels are ignored. Since the US federal funds rate is a central variable of the model, its transmission to the oil market implies the existence of transmission channels.

The five variables are tested for cointegration by the Johansen procedure (Johansen, 1991). There are two cointegrating relationships at the 5 per cent level of significance both concerning the trace and the maximum eigenvalue test statistics. The precondition for running a VAR is thereby fulfilled. The specification takes the following form:

$$B_0 X_t = C + \sum_{i=1}^{2} A_i X_{t-i} + \varepsilon_t \qquad (4.44)$$

where X_t is the vector containing the federal funds rate, ffr_t, the oil spot price, $oilpr_t$, non-commercials' net positions, $noncom_t$, the fundamentals ratio $ip\_oilprod_t$, itself containing industrial production and oil production, and the exchange rate variable, $exch_t$. These are the variables that we have just presented. C is a [5x1]-vector of constants, A_i and B_0 are [5x5]-matrices where i denotes the lag. The lag length of 2 is suggested by the Schwarz information criterion (SIC). The Akaike information criterion (AIC) suggests more lags.[3] We decide in favour of parsimony and choose two lags. ε_t is the vector of innovation shocks. It is assumed that ε_t fulfils the conditions of $E(\varepsilon_t) = 0$, that its covariance matrix is positive-semidefinite, and that it does not face autocorrelation.

B_0 allows the innovation shocks of the variables to have an effect on another variable within the same period. This requires an ordering of the variables to determine which variable affects another variable contemporaneously and which one does not. Cholesky ordering is helpful in this place. As argued in the outline of the empirical proceeding, we mainly want to know how the oil price is affected by the other variables. It is therefore influenced by all other variables in the same period but affects other variables only with a lag. The rest of the order is chosen such that the fast reacting variables are impacted by others more immediately, while they influence slowly reacting variables only with lags. These relationships are expressed by means of the structural shocks u_t in the following decomposition matrix $u_t = B_0^{-1} \varepsilon_t$:

$$\begin{bmatrix} u_t^{ffr} \\ u_t^{ip\_oilprod} \\ u_t^{exch} \\ u_t^{noncom} \\ u_t^{oilpr} \end{bmatrix} = \begin{bmatrix} a_{11} & 0 & 0 & 0 & 0 \\ a_{21} & a_{22} & 0 & 0 & 0 \\ a_{31} & a_{32} & a_{33} & 0 & 0 \\ a_{41} & a_{42} & a_{43} & a_{44} & 0 \\ a_{51} & a_{52} & a_{53} & a_{54} & a_{55} \end{bmatrix} \begin{bmatrix} \varepsilon_t^{ffr} \\ \varepsilon_t^{ip\_oilprod} \\ \varepsilon_t^{exch} \\ \varepsilon_t^{noncom} \\ \varepsilon_t^{oilpr} \end{bmatrix} \qquad (4.45)$$

The federal funds rate is allowed to have the greatest immediate impact on other variables because it is the only variable that is at least to a great part exogenous.[4] The fundamentals variable is placed second, because it is suggested to react more slowly to changes in the environment than the financial market variable, to wit, $noncom_t$.

The impulse responses of this SVAR are shown in Figure 4.4. The result is not spectacular at all. First of all, the federal funds rate features no significant impact on other variables except a scarce effect on the fundamentals ratio, suggesting a weaker reaction of the oil demand side to an increase in the interest rate, than the supply side. The signs of the impulse responses of the oil price, net positions and the exchange rate are even in the opposite sign of what the theory suggests, even though they are not significant. The significant results are on the one hand the negative relationship between the US dollar strength and the oil price, where causality seems to go from the exchange rate to the oil price. On the other hand, there seems to be a significant but quite small positive two-sided causality between the oil price and net positions. The clearest result arises with regard to the fundamentals variable: the oil price increases in response to an increase in industrial production relative to oil supply.

The second period, from 2009 until 2014
Modelling the impact of unconventional monetary policy is not easy. The advantage in comparison to the federal funds rate is that the effect of asset purchases should immediately be visible in prices and not just gradually, as is probably the case with the policy rate of interest. On the other hand, a change in the federal funds rate is only announced at the end of the FOMC meeting and thus unknown before, even though there are trials to anticipate it. Unconventional measures are announced as well but then executed only step by step. For example, the Fed communicated on 13 September 2012, that it would purchase additional mortgage-backed securities of 40 billion US dollars per month (Fed, 2012). While this information is probably new for market participants, subsequent monthly purchases are not new. Thus, a large part of unconventional policy can almost certainly be anticipated once it is announced. By this argument, it is the shock at the announcement date that should be included in the investigation (see, for instance, Gagnon et al., 2011; Gilchrist & Zakrajsek, 2013). However, it is an open question whether or not asset purchases are anticipated effectively. Assuming that they are perfectly anticipated after the announcement is an implicit assumption of the efficient markets hypothesis that we consider to be too strong an assumption. Only modelling the announcement shock as an exogenous dummy in the SVAR is thus likely to prove insufficient. Moreover, it is not clear how to consider a single announcement. Including

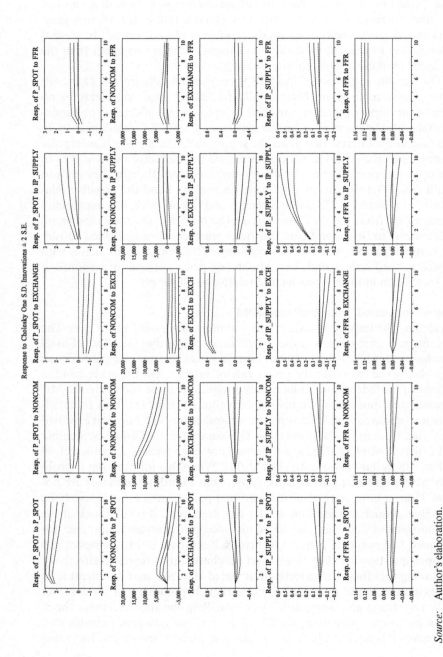

Source: Author's elaboration.

Figure 4.4   Impulse responses between monetary policy and the crude oil market, 2000–08

all dates since the outbreak of the financial crisis where the FOMC signalizes a change in its stance on unconventional monetary policy would require the inclusion of at least 15 dummies (Fed, 2014a). Since the efficient markets hypothesis is hardly able to isolate 15 separate effects, the result may be that competing dummies mutually reduce the chance of significant outcomes when they follow one another with high frequency. It is, in addition, not clear how a dummy variable should be built. It may be only a short-run event or it might leave its trace until the end of the considered time frame.

For these reasons, an alternative approach is chosen. It suffers the inconvenience of being anticipated but qualifies insofar as it can take longer-run effects into account and that it is a set of real data. We take the holdings of the System Open Market Account (SOMA) of the US Federal Reserve. It contains the securities purchased by the Fed in open-market operations. Data are available at the Federal Reserve Bank of New York (2015). It allows drawing a rather exact picture of the central bank's implementation of the asset purchase programmes. The sum of total assets features a very slowly growing and stable pattern from 2003, when the series started being published, until the end of 2008. In 2009, it soon starts rising to levels that are multiples of the initial ones. SOMA changes thus reflect monetary policy changes from the beginning of the financial crisis onwards and especially against the background of a policy rate of interest that has reached the zero lower bound.

The SVAR keeps its form besides two changes. First, the variable $soma_t$ replaces ffr_t. Second, since the central bank now acts not by setting the policy rate of interest but rather by directly intervening in financial markets, we add a stock market variable reflecting financial market conditions. We argue that higher stock prices spill over to commodity markets owing to investors' purposes of profit, wealth store or portfolio diversification. It makes the model less parsimonious but has the potential to test an essential aspect of the theory, namely how monetary policy practically transmits to the oil price. The corresponding data series is the NASDAQ Composite Index (2015) containing more than 3000 equities either in the form of stocks or private non-derivative securities. For simplicity, we refer to it as p_stock_t. It is placed between $exchange_t$ and ip_oils_t in the Cholesky decomposition. The Johansen cointegration test suggests three and one cointegrating relationships at the 5 per cent level for the trace and the maximum eigenvalue statistics, respectively.

The result of this second SVAR is shown in Figure 4.5. It is much more promising in what concerns significance. The SOMA variable has small and scarcely significant but lasting impacts on the oil price, on net positions and on stock prices. This corresponds to the preceding analysis:

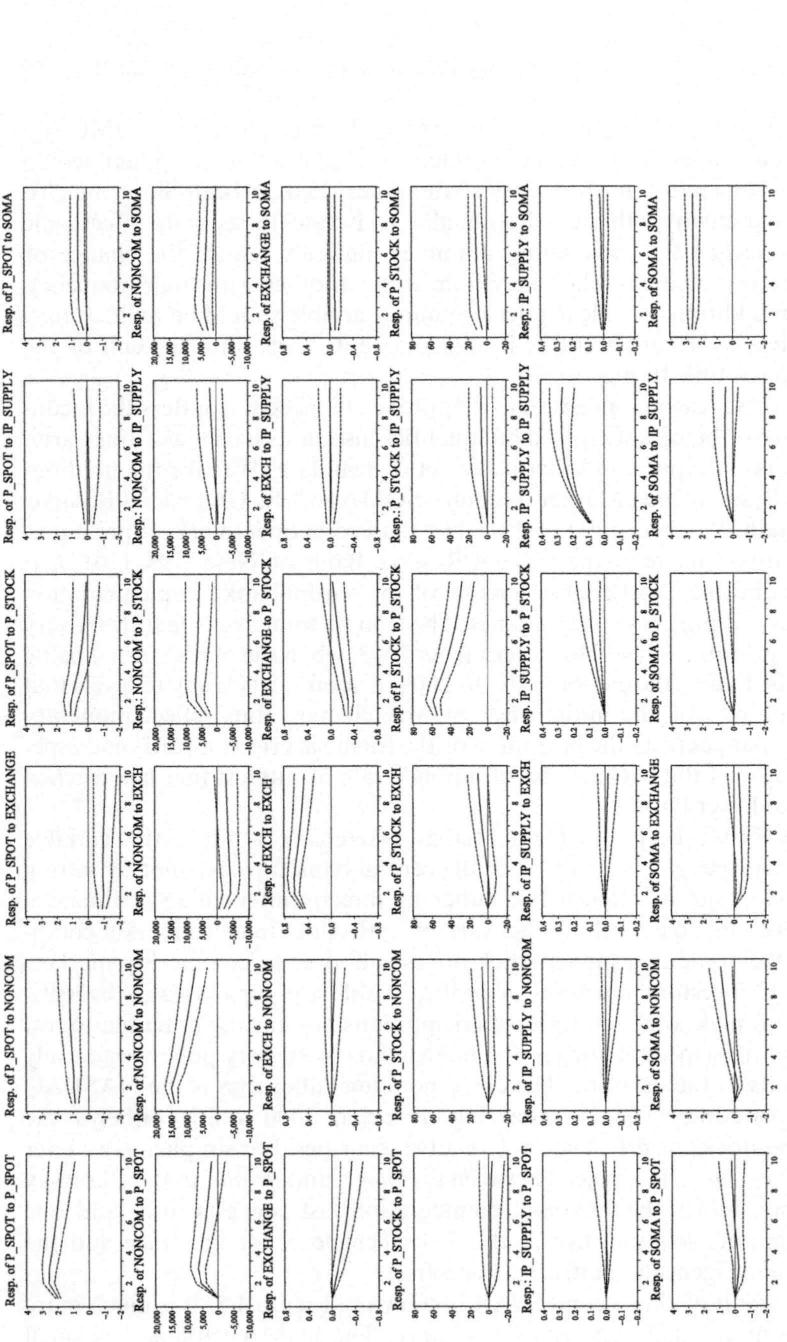

Figure 4.5 Impulse responses between monetary policy and the crude oil market, 2009–14

Source: Author's elaboration.

asset purchases raise the oil price, because investors react by purchasing stocks and futures contracts. It is remarkable insofar as it is hard to find an appropriate monetary policy variable and *soma$_t$* is not a perfect one, either. Moreover, there seems to be a two-sided positive causality between the oil price and non-commercial net positions. Higher net positions, that is, more financial investment, raise the oil price, whereby a higher oil price gives an incentive to raise investment in futures contracts. These effects are confirmed by the significant and positive impact that rising stock prices have on net positions and the oil price. While the effect on the oil price occurs only in the short term, the effect on net positions remains positive for the ten-week window considered. This might be due to the fact that the spillover from stock markets affects net positions directly and thereby the oil price indirectly. The impact of an exchange rate change is about the same as between 2000 and 2008.

The fundamentals variable does not react significantly to any other variable and affects only stock prices. Intuitively, higher industrial production leads the stock market index upwards. The relative passivity of this variable in reacting to other variable changes might partially be due to the need for interpolating monthly to weekly data and partially to more lagging changes in fundamentals than in financial variables.

While one VAR hardly exhibits any significant result, the other is rather well in line with our theoretical analysis. Yet, the latter cannot belie that the VARs have some weaknesses. Cholesky decomposition has been chosen and the resulting ordering is plausible. However, a change in the ordering or another structural decomposition might make even strongly significant impulse responses insignificant. There is thus the possible problem of missing robustness. But, additionally, there is another, probably much more serious criticism: our theoretical analysis draws a picture of the oil market consisting of a spot and a futures market both evolving simultaneously. There are complex relationships that materialize in a different degree and take more or less time to occur. Our theoretical analysis takes radical indeterminacy into account that capitalist economies are confronted with. VARs require summarizing a complex system to some few standardized variables. A fatal drawback of VAR models hardly ever discussed in the literature can be summarized as follows: all variables are obliged to have the same frequency and the same number of lags. Hence, effects of oil market fundamentals are assumed to have the same speed as an exchange rate change or an interest rate change. If not, they do not exist. However, it is obvious that financial variables are more flexible than real variables. We argue that a change in the US federal funds rate has an effect on the oil market that is hard to measure, because it might be anticipated and probably occurs not in the form of a shock but rather gradually. It thereby

becomes quite difficult for the US federal funds rate to prove its importance if the number of lags is given exogenously from the perspective of a
single variable, because it is actually determined by the model as a whole.

4.4.2 A Cointegrating Relationship of the Crude Oil Market

For the reasons summarized in the previous section, we must find a way to
test our theoretical analysis while allowing for indeterminacy and complexity. Let us start with a cointegrating equation over both periods from 2000
until 2014, containing the variables that we argued to be the central ones.
We leave monetary policy aside for a moment and merely consider the
crude oil market separately. The oil price is represented as explained by the
fundamentals, to wit, supply and demand, non-commercials' net positions
in the futures market and the exchange rate. Since the need for parsimony
partially loses its urgency, supply and demand components are taken separately as $oilprod_t$ and ip_t respectively. The basic equation is as follows:

$$oilpr_t = \beta_0 + \beta_1 * noncom_t + \beta_2 * ip_t + \beta_3 * oilprod_t + \beta_4 * exch_t + \varepsilon_t \qquad (4.46)$$

Estimation of equation 4.46 yields results that are shown in the first column
of Table 4.1. To base the starting point of the following proceeding on
stable grounds, it is as well estimated in logs. For this purpose, net positions
have to be transformed, because log values can only be built with positive
numbers. As a simple solution, we index the series with the first data point
being set to 100. Then, we linearly transform it by adding the lowest value to
each data point such that all values are at least zero. Finally, we raise data by
1 in order to have only positive values. This approach may appear as arbitrary but it is the simplest one possible. To broaden the basis of equation
4.46 further, we replace data of $noncom_t$ by another indicator: instead of
net positions of non-commercials, we take their total long positions. The so-
called spread, that is, the number of non-commercial long positions being
evened up by their own short positions, is included in the dataset. Now,
as we argued, net positions might be an insufficient variable to cover all
potential speculative effects. Total long positions are thus a trial to approach
speculative activity from another viewpoint. The higher the long positions –
including the spread – the higher the demand power is exerted in the futures
market. The result should be a higher price. Likewise, we conduct the same
estimation in log values whereby a transformation of data is not necessary
anymore, as they are already positive. We thus end up with four equations.

It can be seen that the most clear-cut significant results are given by the
spot demand side, that is, industrial production and the exchange rate.
Both coincide with economic theory. The higher the industrial production,

Table 4.1   *The oil price explained by fundamentals and financial variables,*
           2000–14

	1) net positions	2) net positions, logs	3) total long positions	4) total long positions, logs
constant	266.43***	10.53***	277.52***	10.90***
	(20.76)	(1.20)	(20.24)	(1.19)
ip	1.69***	1.60***	1.40***	1.32***
	(0.13)	(0.16)	(0.16)	(0.17)
oilprod	−1.70***	0.41	−1.62***	0.29
	(0.22)	(0.30)	(0.22)	(0.30)
exch	−1.77***	−3.33***	−1.74***	−3.17***
	(0.08)	(0.11)	(0.07)	(0.12)
noncom	7.80E–06	−0.01	1.33E–05***	0.05***
	(5.48E–06)	(0.18)	(4.02E–06)	(0.02)
R^2	0.90	0.94	0.90	0.94
DW statistic	0.068	0.11	0.07	0.10

Notes: Standard errors are in parentheses. Coefficients with *, ** or *** are significant at the 10%, 5% or 1% level.

Source: Author's elaboration.

the higher the oil price. The exchange rate variable is negatively correlated with the oil price. The stronger the US dollar, the lower the oil price. The oil supply variable is significantly negatively correlated to the oil price, indicating that higher oil production lowers the price of oil. However, this applies only to estimations (1) and (3): the log estimates are positive but insignificant. The financial market variable, $noncom_t$, is only significant when it represents total non-commercial long positions.

Despite some hints about causal relationships, the regressions in Table 4.1 are correlations that do not provide reliable information about how variables cause one another. Tests for cointegration show that all regressions consist of cointegrating variables. This can be seen from the Augmented Dickey–Fuller (ADF) tests applied to the residuals shown in Table 4.2.

Yet, Table 4.1 shows that the regressions feature impressive R^2s but as well strong autocorrelation exhibited by the Durbin-Watson statistics. Being aware that any correlation cannot stand for itself but rather needs an economic background, this should not disturb our further analysis. It may well be that autocorrelated error terms are due to behavioural anomalies and indeterminacy that may produce short-term disturbances. They do not necessarily falsify our investigation. At any rate, autocorrelated error

Table 4.2   ADF test statistics for cointegration residuals

	1) net positions	2) net positions, logs	3) total long positions	4) total long positions, logs
t-statistic	−4.04***	−5.25***	−4.16***	−4.95***
p-values	(0.00)	(0.00)	(0.00)	(0.00)

Notes: Coefficients with *, ** or *** are significant at the 10%, 5% or 1% level.

Source: Author's elaboration.

terms raise the likelihood that the regression model is wrongly specified. The four regressions are modified by introducing a moving average (MA) component (see Enders, 2014, pp. 50–51; Hamilton, 1994, pp. 48–52). This means that the dependent variable also depends on the error terms of past periods. Potential anomalies can be eliminated by correcting current outcomes by past ones. The result is shown in Table 4.3. Including residuals lagged by two periods suffices to eliminate autocorrelation.[5] R^2s approximate one. Note that the estimated coefficients do not greatly change except that they have become even more significant, now at the 1 per cent level for almost all variables. The financial market variable has become significant for all four regressions. Though, in the second column of Table 4.3 it is significantly negative. Let us point out that transformation of net positions to log data may reduce the quality of the data series and make the estimate less credible. The sign of the supply variable is significantly negative, if normal values are taken and significantly positive for log values. ip_t and $exch_t$ are still significantly positive and negative, respectively.

These regression results are useful to our further examination. All in all, we can argue that most results are rather stable whether taken as logs or not. The financial market variable and oil production require further analysis. Moreover, the coefficients are almost identical in the basic correlation in comparison to the MA representation.

Assessing causalities

The next step of our analysis focuses on causal relationships. We are interested to know if the financial market variable, oil demand, oil supply, and the exchange rate cause changes in the oil price. The method applied is the concept of Granger causality (see, for instance, Granger, 2004, p. 425). By investigating relationships between two variables separately, the number of lags can be varied in contrast to the VARs, where the lags of a particular variable are predetermined by the whole model. The number of lags for measuring Granger causality is not definitely given. Often, it is chosen by

Table 4.3   The oil price explained by fundamentals and financial variables in a moving average representation, 2000–14

	1) net positions	2) net positions, logs	3) total long positions	4) total long positions, logs
constant	278.87***	11.11***	287.83***	11.46***
	(5.39)	(0.38)	(5.28)	(0.37)
ip	1.66***	1.57***	1.37***	1.30***
	(0.03)	(0.05)	(0.04)	(0.05)
oilprod	−1.76***	0.36***	−1.67***	0.21**
	(0.06)	(0.10)	(0.06)	(0.09)
exch	−1.81***	−3.37***	−1.77***	−3.2***
	(0.02)	(0.04)	(0.02)	(0.04)
noncom	9.81E–06***	−0.01***	1.41E−05***	0.06***
	(1.41E–06)	(0.00)	(1.04E–06)	(0.01)
resid(−1)	1.16***	1.12***	1.16***	1.11***
	(0.04)	(0.04)	(0.04)	(0.04)
resid(−2)	−0.19***	−0.18***	−0.20***	−0.16***
	(0.04)	(0.04)	(0.04)	(0.04)
R²	0.99	0.99	0.99	0.99
DW statistic	1.996	1.96	2.00	1.97

Notes: Standard errors are in parentheses. Coefficients with *, ** or *** are significant at the 10%, 5% or 1% level.

Source: Author's elaboration.

the number of lags that a corresponding VAR would suggest by applying information criteria. But this is not an exclusive way of proceeding. Finally, the lags should be chosen according to economic reasoning. Fundamentals variables are expected to require more time to have significant effects on other variables than financial variables.

Taking one single number of lags as the true and only one would be arbitrary. To get stable and credible results, we will investigate Granger causalities at different time lags. The test takes the following general form. If it is to be discovered whether a causes b, the formula is as follows:

$$b_t = \sum_{i=1}^{I} \beta_i * b_{t-i} + \sum_{i=1}^{I} \alpha_i * a_{t-i} + \varepsilon_t \qquad (4.47)$$

where I is the total number of lags included. Equation 4.47 is held in rudimentary form. It does not indicate whether the variables are in levels or in differences. Granger causality testing requires data to be stationary (Hendry, 2004, p. 205). To test the time series for their order of integration we refer to the ADF test (see, for example, Enders, 2014, pp. 206–208). It shows that all variables are I(1), that is, they are unit root in levels and stationary in first differences.[6] Hence, Granger causality is tested in first differences. Equation 4.47 is completed by a level variable of a and b for the first lag. The rest remains the same. This is a usual way of proceeding in practice that is analogous to the formula of the ADF test and does not change the intuition of the equation.

The general procedure when testing for Granger causality is the following: first, to get robust results, the causality test is done with different lag lengths from one to 20 weeks. Second, the sum of the estimated coefficients is considered, that is, the sum of the α_i's in equation 4.47. We not only want to know if Granger causality is given but also whether its sign is in accordance with theory. Estimation is done both with normal and log data. Hence, we have twice a maximum lag number of up to 20, which yields 40 estimates in total. We make a definition how to judge estimate results: if both normal and log data are significant for at least three different lag lengths at the 5 per cent level we say that the causal relationship is stable. If only one of the two data sets reveals significant results but for at least five different lag numbers, we say as well that the connection is robust. Additionally, we have to see whether the sum of coefficients corresponds to what we expect from theory. In the case where there is only little significance, but the sum of coefficients is overwhelmingly in line with theory, we may say that there is at least a partial relationship between the variables of interest. In contrast, if significance is strong but the sign of the suggested effect is in contrast to any economic intuition, then we have to conclude that there is no reliable causal relationship.

The results of the Granger causality tests are quite extensive and thus only briefly summarized in Table 4.4. As they are only of secondary importance, Durbin-Watson statistics are not shown here. Yet, autocorrelation hardly occurs in these tests neither with few nor with many lags. First, as we expected, growing industrial production Granger causes a higher oil price. Second, oil production brings much less distinct results. The tests do not allow for the conclusion that oil production has a significant impact on the oil price. Possible reasons for this shortcoming are emphasized later. The third variable, the exchange rate, clearly affects the crude oil price in the sense of what we expected, namely that a weaker US dollar raises the price of oil. Finally, the test for Granger causality from the financial market variable to the price is once conducted with non-commercials' net long positions and once with their total long positions. Neither outcomes confirm the suggestion that speculative activity raises the price. In

Table 4.4 Granger causality tests for oil market data, 2000–14

Granger causality test from a to b	Significance	Sum of coefficients in line with theory	Causal relationship
ip → *oilpr*	yes	yes	strong
oilprod → *oilpr*	no	partially	no
exchange → *oilpr*	yes	yes	strong
noncom → *oilpr*			
net positions	no	no	no
total long positions	yes	no	no

Source: Author's elaboration.

contrast, sometimes the sum of coefficients is even negative. This would mean that futures market investment of non-commercials lowers the oil price, which is with a high probability economic nonsense.

A preliminary interpretation of the speculation test results
Our findings may provide satisfaction to those who deny any influence of speculation on the oil price. For instance, the *Interim Report on Crude Oil* of the ITF (2008), Stoll and Whaley (2010), Alquist and Gervais (2011) as well as Büyükşahin and Harris (2011) apply similar Granger causality tests normally using net long positions to represent speculation. Even though these studies partially differ in the commodity and investor group considered, they draw a generally common conclusion: first, the oil price (or other commodity prices) has a causal effect on futures price investment, to wit, a higher oil price induces investors to raise their positions. Second, the volume of futures market investment does not affect the oil price, that is to say, Granger causality tests yield, with hardly any exception, insignificant results. There is no reason to criticize the first finding. Logically and as already discussed, many financial investors are trend followers.

The analysis presented here has three advantages in comparison to the studies mentioned above: it investigates two data series that should approximate speculation, rather than only one; it is a larger approach as it includes many lag choices; and it takes the sum of coefficients into account instead of only considering *p*-values. The implicit assumption that the above-named authors probably make is that the effect of futures market investment is around zero, perhaps also slightly positive, but is not significant. However, as mentioned, our Granger causality test results yield sometimes even negative sums of coefficients.

Hence, is it correct to draw from these results the conclusion that speculation does not have any impact on the oil price? There are two possible interpretations. First of all, one may have full confidence in Granger causality measurement techniques. In this case, it is just consequent to deny any effect of futures market speculation on the oil price. However, this interpretation requires an explanation of why the potential effects of the speculative variables on the oil price – whether significant or not – are in large parts negative rather than positive. There is no intuitive way to make sense out of this. A second way of dealing with unfamiliar results is thinking about the weaknesses of the estimation procedure. While the Granger causality test has the advantage that it allows the variation of lags of a single variable in contrast to VARs, it does not incorporate other control variables, which impedes the assessment of isolated effects. This makes again clear that econometric test results can only be interpreted in connection with an economic theory. They cannot provide absolute truth but merely an additional argument in favour of or against a theory.

In this framework, a third possible explanation arises. The issue of speculation is not easy to define and assess. It has been outlined how the futures market works: expectations of agents are heterogeneous and the number of futures contracts is basically unlimited. Long and short positions are two sides of the same coin: they are necessarily always equal. Yet, supply and demand pressures, entering the market from various sides, assert their effects on the futures price. For example, high demand of financial investors may raise their long positions. They may either be offset by the short positions of commercials but also by those of other speculators. Net long positions of financial investors thus probably represent only a part of speculative reality and so do their total long positions. Indeed, negative net long positions, that is, positive net short positions of non-commercials, may potentially be a sign that the oil price should fall, because there is no one to drive up the price by purchasing long positions in large amounts (in this case, however, it should be explained why commercials consequently go net long; probably because they aim at hedging their sales returns against both low and high price extremes). But, contrastingly, it might also be that negative net long positions lead to a higher oil price: for instance, this may occur after financial investors have purchased large amounts of long positions leading up the price. Given that the price moves from a specific level further upwards, a growing number of financial investors may expect the price to fall again at some moment. They start offering short positions to the other speculators. If this continues, non-commercials might end up with net short positions. The consideration of only net positions may give rise to the expectation of a lower price, but it does not take into account the underlying demand pressure that drives the price up. This effect may potentially take place at all levels of non-commercials' participation in futures trade. Whether financial investors' long positions are evened up by other investors' short positions at a high or a low volume does not give a clear-cut indication of what the futures price should be. A particular number of contracts may come into existence by pressure either from the supply side or from the demand side. Assume, first, that some speculators expect the oil price to increase so that they start purchasing long positions or agreeing on new contracts on the long side. These contracts may be offset by other speculators with different expectations. The exerted demand pressure drives the oil price up; futures supply by other investors' offer of short positions is only a reaction. Assume the second case, where the same market participants expect a falling price and hence purchase short positions. Offsetting by other investors may lead to the same volume of futures contracts. However, since the origin of the futures contracts is in this case to be found in supply pressure rather than in demand pressure, the oil price is likely to fall.

All these stylized cases of what form speculation can take are likely to occur simultaneously in the futures market. Moreover, one single effect is possibly quite short-lived and soon replaced by another one. It is thus the high elasticity and uncertainty of the futures market and of financial markets in general that makes it extremely difficult, if not impossible, to represent speculation by a single variable. Considerations suggest that speculation is too complex to be modelled in this way.

Of the three interpretations of the relationship between futures market investment and the oil price, the second one and specifically the third one are more plausible than the first one, which implies unconditional belief in econometric estimation results. They ask for other approaches in testing for speculation. In this respect, it is wrong to conclude that if Granger causality tests yield insignificant results for the causality from net positions to the oil price, then there is no speculative effect in the oil market. The mentioned papers therefore should not be seen as a final verdict proving the non-existence, or non-effectiveness, of speculation.

After the criticism of these approaches, an explanation of the SVARs' impulse responses in Figures 4.4 and 4.5 is needed. In the first period, from 2000 until 2008, the effect of net positions on the oil price is almost insignificant, while it is significantly positive in the second period, from 2009 until 2014. We are now cautious about delivering an interpretation. The result may indeed be an indication of speculation, saying that it was ineffective in the first period and effective in the second period. Thanks to the SVAR approach, which allows isolating individual effects, the effect is as it should be, that is, positive rather than negative. This would also mean that speculation has changed over time and that it is more than one-dimensional such that net positions as a single data series are not sufficient to describe it. Hence, owing to uncertainty, the result is rather unstable. We thus need an alternative approach.

By the way, the problem of representing speculation does not mean that the variables employed are completely useless. On the one hand, it is especially the pattern of total non-commercials' long positions that shows how financialization has increased in the past. We argued that the latter contributes positively to the influence of speculation. Correlations thus can help to give at least a hint of some tendencies. On the other hand, the cointegrating relationship assigns a certain explanatory power to both variables even though the sign is in line with our theoretical analysis only in six out of eight estimates. Yet, what is still missing is the direction of causalities between the variables.

4.4.3   Assessing Monetary Policy Transmission to the Oil Market

In the previous section, the cointegrating equation 4.46 and the subsequent tests for Granger causality have focused on the crude oil market itself without taking monetary policy into account. Before we take a further step to enlighten the speculation issue, let us consider that the oil market is connected to monetary policy. We will point out that the influence of speculation can be assessed or at least be approximated by the analysis of the transmission of monetary policy through financial markets.

To arrive there, we investigate the impact of monetary policy on each of the variables in the cointegrating equation. Testing for Granger causality from monetary policy to the oil price means measuring the total effect, neither distinguishing between fundamentals and financial market effects nor taking any special features of the individual transmission channels into account. Examining the connection between monetary policy and the explaining variables, that is, industrial production, oil supply, the US dollar exchange rate and the speculative variables, then should allow for conclusions about how monetary policy transmits through fundamentals and in how far financial markets play a role.

While the cointegrating relationship of the crude oil market can be estimated for the whole time frame from 2000 until 2014, monetary policy again needs to be separated into two periods. We use again the federal funds rate for the period from 2000 until 2008 and the SOMA variable for the second phase, from 2009 until 2014. Note that the approach of varying lag choices brings the advantage of reducing measurement problems of monetary policy shocks. The more lags are included, the higher the probability that the gradual effects of the US federal funds rate are taken into account.

Again, variables have to be tested for stationarity. As a summary, unit root test results indicate that all variables are I(1) in the first period except industrial production and net positions which are I(2) and I(0), respectively. For industrial production, the variable has therefore to be taken in second differences. In the second period, orders of integration are I(1) at the 10 per cent level for the SOMA variable and at the 1 per cent level for all other variables.

The effect of the federal funds rate in the first period

The effects of monetary policy on the crude oil market variables in the first period are exhibited in Table 4.5. An overall effect of a change in the federal funds rate on the crude oil price cannot be shown. More than that, and what is not shown here, the test results suggest even a positive relationship for some lags which implies that restrictive monetary policy seems to raise the price. The same counterintuitive result arises from the causality test from monetary policy to industrial production. Oil production does not react significantly to a change in the federal funds rate. As mentioned, the oil supply side is investigated in a subsequent section. There are plausible reasons for arguing that oil production is a variable featuring a rather specific behaviour. The next variable, the US dollar exchange rate, reveals ambiguous results. There is high significance for most lag numbers but the sum of coefficients is often negative such that a higher interest rate causes a depreciation. This is not implausible because we argue that in a specific situation, expansive monetary policy may be a sign of financial stress, which attracts investors to raise demand for US dollars. A great part of federal funds rate changes in this period took place in either a boom or a crash phase. It has been suggested above that in such moments, monetary policy is less powerful or achieves the opposite of its objective. For a

Table 4.5   *Granger causality from monetary policy to the oil market, 2000–08*

Granger causality test from a to b	Significance	Sum of coefficients in line with theory	Causal relationship
ffr → *oilpr*	yes	no	no
ffr → *ip*	yes	no	no
ffr → *oilprod*	no	no	no
ffr → *exchange*	yes	partially	strong
ffr → *noncom*			
net positions	yes	partially	weak
total long positions	yes	partially	weak

Source: Author's elaboration.

longer-run negative positive, we rely on existing literature that suggests the depreciating effect of monetary policy (see, for instance, Eichenbaum & Evans, 1995). Even though the hitherto applied futures market variables have been criticized above, they are used again to gain potential hints. This time, performance is improved a little, but if we can conclude that a lower federal funds rate causes higher speculative activity, the relationship is only very weak.

The observation that a higher US federal funds rate does not directly lead to a lower oil price is likely to be related to the well-known 'price puzzle'. This phenomenon, intensely debated particularly by Sims (1992) and Eichenbaum (1992), denotes the observation that contractionary monetary policy is followed by a higher general price level (Eichenbaum, 1992, p. 1002). There may be strong economic activity leading to increasing prices. Monetary policy reacts by raising interest rates. The cause of price growth does not fully disappear. Nevertheless, prices keep rising, but less than they would if there were no reaction of monetary policy (Sims, 1992, p. 988). The price puzzle is one of the sources of the debate about how to model proper monetary policy shocks. It keeps discussions about the effectiveness of monetary policy alive (see Kuttner & Mosser, 2002, pp. 17–18).

The test for Granger causality from the US federal funds rate to the oil price probably suffers a similar problem. Table 4.4 shows that industrial production as an approximation of oil demand has a positive effect on the oil price. At the same time, industrial production is also not too bad an indicator of economic activity in general. Rising industrial production thus may raise the oil price as well as the general price level. Therefore, monetary policy is likely to raise the interest rate level in situations where the price of oil is high. Since the underlying price driver, that is, industrial production, may nevertheless grow further, the price is likely to keep increasing for a while, too. Note that we do not at all mean by this argument that the general price level and the oil price move in the same way, so that the real oil price remains the same. Inflation and oil price developments may tend to move in the same direction, but there is no reason to assume that they do it by the same amount. It is for this reason that core inflation and overall inflation are distinguished. Moreover, it has been shown that the correlation between the nominal and the real oil price approximates unity, confirming that inflation is not useful for explaining a substantial share of nominal oil price fluctuations.

A second effect in the case of crude oil is that it is not only fundamentals that react sluggishly to changes in the conduct of monetary policy. Many financial investors in the futures market probably adapt their decisions – beyond what concerns ultra-short-run speculative strategies – to an altered interest rate level, step by step after the new rate has proved to be stable for a while. A higher interest rate raises refinancing cost more and more, as credits have to be renewed more and more. The price puzzle in the market for crude oil may thus also be due to the fact that the US federal funds rate works only gradually through the financial market effect as it also does through fundamentals.

Concerning industrial production, the Granger causality tests suffer from an analogy to the price puzzle. The central bank tends to contract monetary conditions in a situation when industrial production, or output in general, feature a rising trend or even a boom. These are periods when inflation rates are usually higher than in normal or recessive times, or when there are fears of growing bubbles and instability in financial markets. Like prices, output tends to rise further for a period of time after the monetary authority has raised the interest rate. The opposite happens in a recession: the central bank cuts the interest rate, aiming at stimulating the economy, but output continues to fall. There are good examples for both cases in the period from 2000 until 2008. From mid-2004 until mid-2006, the US Federal Reserve raised the federal funds rate continuously. The FOMC justified these measures by the potential threat of inflation owing to high capacity utilization in production and rising commodity prices (Fed, 2006). Later, when the financial market downturn started in mid-2007, the US federal funds rate was cut again and again, until it reached the zero lower bound in the second half of 2008. As is well known, this could not prevent the slump in industrial production. These are just two events that might be sufficient to yield significantly positive Granger causality from the US federal funds rate to industrial production. Again, one should mention that the boom and the recession would have been more extreme had the Fed not reacted by changing its federal funds rate target.

Yet, econometric measurement problems cannot hide the probable fact that the influence of monetary policy on output is limited. Even though it has the potential to affect fundamentals variables significantly and permanently without the necessity of bringing them back to some hypothetical equilibrium, monetary policy itself is not always a sufficient condition to set the economic system into motion. We discussed in detail at the beginning that money is demand-determined. Expansive monetary policy is not effective if demand for credit is missing. Monetary policy effects exist but their strength and degree of significance change over time so that they are not linear.[7]

The effect of quantitative easing in the second period
Before we address the critical points of the first-period estimates, we analyze the second time frame of interest. For the period from 2009

until 2014, the federal funds rate is replaced by the Fed asset purchases. Table 4.6 shows the respective Granger causality test results. The significant effect of the SOMA variable on the oil price confirms the results of the SVAR. Yet, it might also suffer from the same kind of pricing puzzle as the estimates of the first period. On the other hand, the positive effect of quantitative easing on the oil price is possibly even underestimated, as it is easy to anticipate asset purchases regularly taking place month after month after they have been announced by the FOMC. We try again to separate the financial market and the fundamental effects in the crude oil market.

Even more than in the first period, Granger causality tests do not reveal causal effects from monetary policy to global industrial production. However, this time the background is different. Quantitative easing measures are applied when the conventional transmission mechanism is broken or if the target interest rate reaches the zero lower bound. The top priority then is to purchase assets in order to bring market interest rates down. The effect of interest rate changes on the real economy comes in a second step. We took this special feature of unconventional monetary policy into account by distinguishing first- and second-stage transmission. The specific recessionary or stagnating environment in which quantitative easing measures are applied might partially be responsible for the absence of a significant effect of the SOMA variable on industrial production. The following test variable, oil production, is still an insignificant variable in connection with monetary policy. The supposed reasons for the determination of oil production are still to be explained. The relationship between monetary policy and the exchange rate is significant and in line with theory. The futures market variables react at least partially significantly to

Table 4.6   Granger causality from monetary policy to the oil market, 2009–14

Granger causality test from a to b	Significance	Sum of coefficients in line with theory	Causal relationship
soma → oilpr	yes	yes	yes
soma → ip	no	yes	no
soma → oilprod	no	yes	no
soma → exchange	yes	yes	strong
soma → noncom			
net positions	yes	yes	yes
total long positions	yes	no	no

Source: Author's elaboration.

expansive monetary policy. However, against the background of the above discussion, the evidence is too weak to be robust.

Bringing significant and insignificant results together

Let us now summarize our findings. With regard to the causalities in the crude oil market, we have evidence of a significantly positive effect of industrial production and a significantly negative effect of the US dollar exchange rate on the oil price. Oil production and the futures market variables do not provide conclusive results. The estimates for the period of conventional monetary policy suffer price puzzle problems. The overall effect of the US federal funds rate of interest on the oil price is found to be overwhelmingly positive instead of being negative. The interest rate impact on industrial production and oil production is insignificant, so that it becomes hard to detect monetary policy transmission through fundamentals in accordance with the ambiguity suggested by our theoretical investigation. The exchange rate of the US dollar reacts to a change in the US federal funds rate partially in accordance with our theoretical analysis, while the speculative variables still are not helpful for yielding intuitive outcomes. The second period gives a similar picture. The effect of SOMA on the oil price is positive, but its validity may again be limited by the price puzzle. Both fundamentals variables are still insignificant and so are futures position data. The only significant variable is the exchange rate.

Even though these results seem to be rather disappointing, we can read more out of them than we might think at first sight. The summary of the hitherto results allows us to find a way of proceeding further. We noticed that growing industrial production raises the oil price. Therefore, the significant Granger causality tests between these two variables lead us to conclude that the significant coefficient in the cointegrating equation 4.46 is probably not just a correlation but in fact represents a causal relationship. Moreover, in the same regression, the coefficient estimate for the exchange rate also seems to reflect a causal effect. We have thus two variables that can explain at least a fraction of the oil price development. The oil supply variable is still to be explained. Yet, oil production is found to be significant in the cointegrating equation for normal data. As it is a fundamentals variable, it is strongly connected to the demand variable, that is, industrial production.

Let us therefore simulate a 'fundamentals component' of the oil price. It is the price that would result if it were only explained by the estimated coefficients of industrial production, oil production and exchange rate variables. It may be asked why the exchange rate is taken as a fundamentals variable. Following the distinction of monetary policy transmission channels above, the exchange rate channel of monetary policy transmission is divided into a fundamental and a financial market component. The latter implies futures market reactions to changes in the exchange rate that are themselves due to monetary policy. The way in which the exchange rate variable is employed now follows the reasoning of how it is effective through fundamentals. Without ruling out that it may also contain speculative aspects, we use it, for convenience, as a fundamentals variable.

The fundamental price of oil as defined now leaves a residual, which is the difference between the true price and the constructed price. Given that the fundamentals variables are reliable and more or less sufficient to cover all important effective movements in fundamentals, the residual should represent the financial market component of the oil price. By measuring the causal effect of monetary policy on this residual, we may find out if there is a significant policy transmission through the futures market. As a concession, however, the assumption that the fundamentals and the financial market component of the oil price are numerically separable has to be accepted for this test. ADF tests of the speculative components, p_{fin}, implied by the variants of the cointegrating equation 4.46 with normal data show that the variables are stationary at the 5 per cent level of significance in both periods. To have a stronger guarantee for stationarity, they are nevertheless taken in first differences where unit root is rejected.

The result for the first period is plotted in Figure 4.6. There are two graphs in each panel (a) and (b). One graph shows the speculative component of the cointegrating equation with non-commercials' net positions, while the other graph shows the one with total long positions. Log estimates cannot be realized, since the financial market component contains the original residuals ε_t of the cointegrating equation. They are both equally positive and negative owing to their unbiasedness. Transformation to positive numbers bears the risk of unpredictable measurement mistakes. Hence, we take variants (1) and (3) of the cointegrating regression for the following procedure.

The US federal funds rate of interest has a significant effect on the financial market component of the oil price for small lag lengths as well as for high ones from 14 upwards, as can be seen in panel (a) of Figure 4.6. In between, p-values move far away from significance levels. Both graphs are almost identical. This is due to the fact that the coefficients of the fundamentals variables are rather similar in both equations. Moving from panel (a) to panel (b), we see that the sums of coefficients are almost perfectly in line with economic theory. While they are positive for small lag numbers, they decrease becoming clearly negative very soon and decreasing further almost steadily. As a reasonable interpretation, the causal relationship between the US federal funds rate and the financial market component of the oil price suffers a kind of price puzzle for small lag choices, which drives

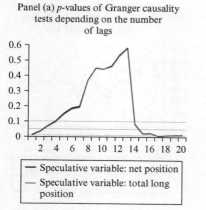
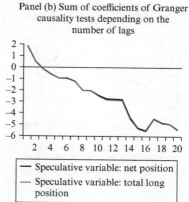

Source: Author's elaboration.

Figure 4.6   *Granger causality from the US federal funds rate to the*
              speculative component of the oil price, 2000–08

the sums of coefficients positive. With higher lag numbers, the price puzzle is overcome and the relationship becomes more and more negative. It is also possible to bring this interpretation in accordance with the *p*-values in panel (a). Low lag choices bring significantly positive results. Increasing the number of lags lets the sum of coefficients approach the zero line in panel (b). Logically, this yields more insignificant results, since the F-test statistic of the Granger causality test cannot reject the null hypothesis of all coefficients being equal to zero. As long as the sums of coefficients move around zero, the *p*-values are above significance levels. When the former are more negative, causality estimates again become significant.

While it was clear from the beginning that monetary policy transmission to the oil price through fundamentals is ambiguous, our theoretical analysis predicts transmission through financial markets to be clear in the sense that expansive monetary policy raises the oil price. By filtering out the fundamentals component in an imperfect but reasonable way we now see that the financial market component behaves as suggested in our theoretical analysis. Lower interest rates seem to fuel speculative activity, which raises the oil price on condition that fundamentals of the oil market remain unchanged. Intuitively, a change in the interest rate level appears to materialize gradually. This is confirmed by the finding that the negative relationship between the US federal funds rate of interest and the speculative component of the oil price strengthens with an increase in the number of lags.

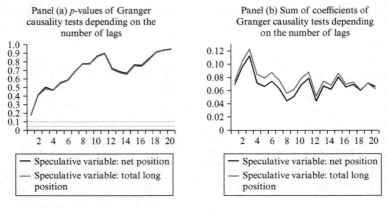

Panel (a) *p*-values of Granger causality tests depending on the number of lags

Panel (b) Sum of coefficients of Granger causality tests depending on the number of lags

— Speculative variable: net position
— Speculative variable: total long position

Source: Author's elaboration.

Figure 4.7   Granger causality from SOMA to the speculative component of the oil price, 2009–14

In the second period, from 2009 until 2014, the same Granger causality test with the SOMA variable yields only insignificant results as shown in panel (a) of Figure 4.7. Panel (b) shows at least that the sum of coefficients is always positive and rather stable. Hence, it is not only positive for the oil price as a whole but also for the speculative component of that price. Yet, insignificance does not allow for drawing a conclusion. An important reason for this might be the strong anticipation of asset purchases by financial investors after the US Federal Reserve announced them. If so, it is quite hard to find significant results.

There are approaches to calculating a 'shadow' federal funds rate of interest. After the policy rate of interest reached the zero lower bound, quantitative easing had the task of depressing market rates of interest further. Thus, unconventional monetary policy can be considered as a suitable measure to pull the US federal funds rate of interest virtually below zero. Wu and Xia (2014) provide a dataset of such a shadow rate. However, it cannot but be developed out of a model that itself requires strong assumptions. Moreover, it only exists at monthly frequency. Granger causality thus can hardly yield any meaningful results, because all shadow US federal funds rates of interest are constructed issues and highly artificial. Therefore, we renounce making any further tests.

After these different tests, we have a result that partially confirms our theoretical suggestions. While significance is impressive for the first period, it is missing for the second one. We must rely on intuition, which tells us that quantitative easing also has a positive impact on the oil price but through ways that are less linear and therefore more difficult to detect. On the other hand, the SVAR of the second period assigns a more significant effect on the oil price and speculative activity. Even though we criticize this approach and the speculative variables, it cannot be said that there is no empirical evidence of monetary policy effects through the futures market in the period from 2009 until 2014. If there were a single variable able to represent speculation, it would account for a larger share in the cointegrating equation. The financial market component of the oil price would be larger. Hence, one may argue that the current result even underestimates the influence of speculation and hence the influence of monetary policy.

4.4.4   What Crude Oil Inventories Can Tell Us

Another way to assess price-influencing speculation is the investigation of oil inventories. This topic is referred to so often that it deserves being considered more closely. As discussed in our theoretical analysis, many economists take the inventory argument as a proof of whether speculation exists or not (see, for instance, Alquist & Gervais, 2011; Frankel, 2014; Hamilton, 2009; Krugman, 2008). By stating that inventories did not rise extraordinarily when the oil price peaked in 2008, it is frequently argued that there was no speculation. There are, however, many reasons to criticize this view. In this section, a new approach is presented. It may show how the inventory argument can nevertheless be used productively.

We have found some main weak points in the argument: there are problems to get appropriate data, since the latter probably include also strategic reserves. Moreover, underground reserves should as well be counted as reserves, but it is not clear to what extent. In addition, the assumption that capacity utilization is constant is rather unrealistic. And finally, precautionary demand in the spot market may lower inventory measures but lead to a higher oil price all the same.

Let us start with the latter issue by using the SFC model. To recall, oil demand in the spot market is given by the following equation:

$$C_{oil,d} = \delta_0 + \delta_1 {}^* C_s - \delta_2 {}^* p_{spot} \tag{4.1}$$

The coefficient δ_2 determines how spot demand reacts to an increase in the spot price. As we argued, the slope of the spot demand curve is likely to be negative over the medium and long run. Speculation in the futures market raising the oil price then requires higher inventories owing to lower demand in the spot market. Measuring inventories and concluding whether there is speculation or not is impeded by uncertainty. At least in the short run – however long the short run might last – the demand curve slope can also be positive. Consumers observe a rising price and hence expect it to rise even higher in the future. As a reaction, they raise demand for oil now in order to save an amount of money equal to the difference between the expected price in the future and the price today. Inventories measured in oil companies' stocks then decrease even though speculation raises the price away from fundamental conditions. We do not know whether such precautionary demand is motivated by speculative or hedging purposes.

In panel (a) of Figure 4.8, there is the pattern of inventory accumulation in the baseline simulation of the SFC model where the interest rate target is cut by the central bank. The reaction of the markets has been explained above (see Figures 4.1 and 4.2). Growing speculation raises inventories in the spot market. When the oil price falls below its initial value owing to real oil industry investment, inventories decrease slightly, because they are used to satisfy growing consumption. In this simulation, δ_2 is equal to 1, so that the oil price affects spot demand negatively in equation 4.1. In panel (b), we assume that precautionary demand comes into play for a limited time span, from the beginning of 2001 until the end of 2002. δ_2 takes the value −0.1. Alongside this, behaviour is as shown in panel (a). We see that inventories react quite confusingly to somebody who is not aware of the possibility of precautionary demand. Inventories fall sharply and then rise again in a concave way until they shoot up when δ_2 again becomes positive, which has the effect of a shock.[8] The central feature of Figure 4.8 is that changes in demand behaviour affect inventory data. Uncertainty about the slope of the demand curve thus makes it quite difficult to gain useful insights about the evolution of oil inventories. If δ_2 varies continuously, producing effects like in panel (b), it becomes hard to find significant results.

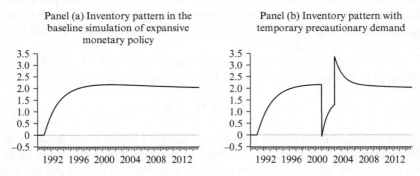

Source: Author's elaboration.

Figure 4.8   *The effect of precautionary demand on oil inventories in the SFC model*

Let us move from the demand to the supply side. The SFC model characterizes the production behaviour of the oil industry in the following way:

$$\Delta IN = \delta_3 * K_{-1} - \gamma * IN_{-1} - C_{oil,\,s} \qquad (4.2)$$

Inventory accumulation of oil producers is simply equal to crude oil produced minus crude oil sold, that is, $C_{oil,s}$. δ_3 represents the production technology, which translates capital input into oil output. The first term on the right-hand side of equation 4.2 thus represents production capacities. If these capacities remain stable or grow over time, if they are fully used and if the spot demand curve is falling, oil inventories necessarily must increase in the presence of a higher spot price that is caused by speculation. The latter condition of demand behaviour has been discussed just before. It is also reasonable for the former condition of at least constant production capacities to be taken as given, since there is no reason why these capacities should decrease when speculation occurs. The second condition of full-capacity utilization may even be relaxed for the consequence of rising inventories still to be valid. They also do so if the rate of capacity utilization is below unity but remains stable over time. Yet, there is no reason why capacity utilization should be stable. The second term of equation 4.2 describes the degree of capacity utilization, which is assumed to depend on the existing level of inventories. γ is positive, so that capacity utilization is below unity whenever inventories are greater than zero. When oil companies have built a volume of stocks sufficient to hedge against reasonable unforeseen disturbances, there is no need for them to accumulate oil stocks further. Effective inventory accumulation can therefore be quite variable over time. Oil production is only bounded above by production capacities.

One may say that this argument does not put Hamilton's (2009) conclusion into question, because lower oil inventories due to lower capacity utilization just imply larger underground stocks of crude oil. This idea is broadly in line with Frankel (1984, 2006) and Anzuini et al. (2013, p. 135). Yet, first, underground inventories can hardly be assessed, because they are not separated from total oil reserves. Second, the authors mentioned above make capacity utilization depend on an arbitrage condition that has been explained and criticized in our theoretical analysis. If so, then over- and underground inventory-building is theoretically tied to a rule of mathematical precision. Our point here is clearly different. In an SFC framework, inventories are built based on entrepreneurial decisions that are modelled in a macroeconomic manner. The fact that oil inventories evolve according to oil companies' feedback resulting from preceding developments means that there is a much wider field open for their pattern. Most likely, oil companies do not leave oil under the ground just because an arbitrage condition tells them to do so. They rather stop accumulating inventories in a situation where they have already enough and they want to save carrying cost. Hence, the feedback coefficient is likely to vary over time, depending on the price level as well as price volatility and suggested risk in the oil market. Moreover, geopolitical events may affect capacity utilization, too. And crucially, we do not limit speculation to the spot market in contrast to the arbitrage condition mentioned above. In that theory, it is oil producers who influence the price by holding back oil instead of selling it. Our model also emphasizes speculation in the futures market, in the presence of which oil producers adapt their inventories. That financial investment in the futures market probably has a greater price-influencing potential than speculation in the spot market has been argued in our theoretical part in Chapter 2.

Figure 4.9 shows how the pattern of oil inventories changes depending on capacity utilization predicted by our SFC model. It is a rather schematic figure, as it takes the feedback coefficient γ to be determined once and for all. In reality, γ changes over time. Inventories then do not inevitably rise in a moment of intensive speculative activity compared to a moment of almost no speculation. Again, temporal variations of

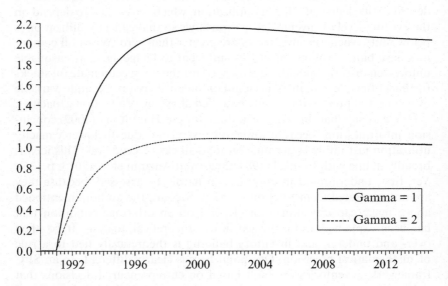

Source: Author's elaboration.

Figure 4.9   The effect of speculation on oil inventories depending on capacity utilization

translating into inventory fluctuations impede the finding of significant results.

Together with equation 4.3 of the model, this argument explains why we found the oil production variable to be insignificant in almost all tests. Oil supply is necessarily always equal to oil demand. Industrial production has been identified as a significant demand force. From the perspective of the crude oil market, it is largely exogenous. Oil supply is therefore more a reaction to oil demand than an independent force. The difference between oil supply and oil production is given by the accumulation of inventories. As we know, inventories can take on a highly uncertain pattern. Oil production thus has hardly the chance to explain a great part of the driving forces of the oil market. It only would have done so on the unrealistic neoclassical assumption of permanently full-capacity utilization. It is only then that production really reflects supply-side constraints.

This argument might be surprising against the background that crude oil is an exhaustible variable. We may expect shifting supply-side constraints owing to limited oil reserves in general or certain oil fields being on the decline in particular. In such a case, inventories should come to their lower limits, so that any further supply shortcoming raises the oil price. Even though reserves are certainly declining over decades, we do not find any systematic significant signs of this tendency in oil production over the short run. Likewise, we do not find oil production to be a meaningful and significant variable. In contrast, oil production seems to accommodate a change in demand (besides short-run capacity constraints like delays in capacity enlargement). This is an indication that the exhaustion of oil reserves is not a binding constraint at this moment.

Any trial to measure how speculation might be associated with crude oil inventories should take these supply- and demand-side concerns into account. The demand-side problem of precautionary demand could be overcome, if there were a measure of inventories that includes not just those of oil producers, traders and other important entities. It should also contain stocks of the firms and corporations of the rest of the economy as well as those of households. However, the EIA as the main provider of stock data includes only primary stocks, to wit, those in refineries, bulk terminals, and pipelines (EIA, 2015d). Inventories outside of the oil industry are not taken into account. Indeed, the problem of precautionary demand is not that consumers start consuming more oil in expectation of a rising price than they would otherwise. Rather, they purchase it for future use and thus also build inventories either in business or privately. The core of the issue is that precautionary demand draws oil out of the official inventory statistics, because it is no longer held by oil producers. There would be a simple way out of this problem by approximating inventories or the change in inventories by the difference between global oil production and global oil consumption. Yet, in analogy to inventory data, consumption data do not cover what is effectively consumed but what is purchased and hence disappears from the oil market. Nevertheless, we use this differential. Despite its imperfection, it still has the advantage that it approximates global and not only OECD inventories. This is in line with equation 4.2 of our SFC model and leads to the following equation:

$$\Delta in = oilprod - oilcons \qquad (4.48)$$

The supply-side aspect of capacity utilization can be taken into account just by investigating how far the utilization rate or the rate of spare capacities change when the crude oil price changes. Empirically, speculation can be said to exist if a higher oil price is associated with either an increase in our inventory measure, a decrease in capacity utilization, or both. In total, we want to know whether the difference between production capacities and oil consumption increases when the oil price rises. This is implicitly contained in equation 4.2. What we want to estimate is therefore given by:

$$oilpr \sim capcity - oilcons \qquad (4.49)$$

Since capacity utilization is a measure of oil production, both variables can theoretically be considered as the same if they are taken in logs. If capacity utilization increases by 1 per cent (where the existing utilization rate is 100 per cent), oil production does likewise (on condition that the production technology has constant returns to scale). By inserting the oil production variable once negatively and once positively into relation 4.49 and then integrating equation 4.48 we get:

$$
\begin{aligned}
oilpr \sim\ & capacity - oilprod + oilprod + oilcons \\
=\ & capacity - utilization + oilprod - oilcons \\
=\ & capacity - utilization + \Delta in \qquad (4.50) \\
=\ & spare + \Delta in \qquad (4.50')
\end{aligned}
$$

Either 4.50 or 4.50' can be estimated, depending on whether total capacities and the utilization rates are investigated separately or summarized in spare capacities (which is just the difference between total capacities and utilization rates). The final equation controls for industrial production, oil production and the US dollar exchange rate. Even though oil production is hardly a helpful variable, we include it as it is significant in the cointegrating equation (4.46) and it may nevertheless add to the overall quality of the estimation model. The regression gets the following forms:

$$oilpr_t = \beta_0 + \beta_1{}^*capacity_t + \beta_2{}^*utilization_t + \beta_3{}^*\Delta in_t + \beta_4{}^*ip_t + \beta_5$$
$$* oilprod_t + \beta_6{}^*exch_t + \varepsilon_t \qquad (4.51)$$

or:

$$oilpr_t = \beta_0 + \beta_1{}^*spare_t + \beta_2{}^*\Delta in_t + \beta_3{}^*ip_t + \beta_4{}^*oilprod_t + \beta_5{}^*exch_t$$
$$+ \varepsilon_t \qquad (4.51')$$

Data are now monthly, since it is the highest frequency available for capacity and global production and consumption data. Capacity data are reflected by an index of drilling wells of the US oil and gas industry as well as their degrees of capacity utilization and spare capacities, respectively, provided by the Board of Governors of the Federal Reserve (2015a, 2015c). Merging both oil and gas data provides imprecise results. But both markets share many similarities, so that test results should nevertheless be useful for interpretation. The dataset does not measure production capacities directly but is an indicator of how capacity evolves. The alternative global inventory variable is based on data of the EIA (2015b). While this inventory measure, that is, production minus consumption, is global, capacity and utilization data cover only the US economy. However, on sufficiently competitive conditions, they allow for some conclusions about the behaviour of the global crude oil market. Moreover, US capacity and utilization data have the advantage in comparison to global data that they exclude OPEC. The latter may behave strategically and thus complicate the interpretation of the results.

Log specification allows for giving variables with small values their appropriate weight. This is particularly relevant for utilization and spare capacity data that are given in percentage points. On the other hand, the alternative measure of inventories is a variable in first differences as it provides the change in stocks in each period. The series thus contains positive as well as negative values. In order to prevent the transformation of the series that potentially jeopardizes important information, we just transform it linearly to an index with the value 100 in the first month. Hence, the series is at least partially adapted numerically to the other series.

Table 4.7 exhibits the estimation results of regression equations 4.51 and 4.51'. The results are quite clear. The rate of capacity utilization decreases significantly when the oil price is high. Likewise, oil inventories grow at the same time. The fact that the coefficient is small compared to the others does not mean anything, since it is the only non-log variable. Moreover, the coefficient does not say anything about the effective historical weight of a variable. A small coefficient can still make a big difference if the changes in the estimated variable are large. Both estimates confirm the presence of speculation: if fundamentals were the only source of oil price changes, oil inventories and capacity utilization should be constant. Or even to the contrary, inventories should decline while capacity utilization should rise in tendency. The oil and gas wells index, briefly denoted as *capacity* in the table, does not change significantly with the oil price. This shows that lower relative capacity utilization is not due to a strong increase in capacities, which would, in relative terms, lead to a drop in capacity utilization, that is, the capacity utilization rate. The result rather suggests that capacity utilization decreases not only in relative but also in absolute terms when the oil price is high. Estimates of regression 4.51' confirm the interpretation as spare capacities increase significantly with the oil price. The control variables yield familiar results with significant estimates for industrial production and the exchange rate and insignificance for oil production. All in all, if there exists the phenomenon of precautionary demand, the results tend to be underestimated. In this case, in fact, inventories grow more than is recognized by data. This is due to the inability of petroleum consumption data to include inventories accumulated in the non-oil industry or in private households. We have already argued above in this sense.

One may argue that both regressions are potentially meaningless, owing to considerable autocorrelation detected by the Durbin-Watson statistic. As in any econometric test, this case is not ruled out. Recall, however, that we are not in search of a significant outcome that would prove that one variable of the regression explains another one. Rather, results in Table 4.7 describe a correlation in its most proper sense. We do not explain the oil price by an increase in oil inventories or a decrease in capacity utilization. Table 4.7 just shows that a higher oil price seems to be associated with lower capacity utilization and inventory growth. This – nothing more and nothing less – is what we suggested theoretically in preparation of these estimations. It is now confirmed empirically. Therefore, despite autocorrelation, our significant results are useful. Moreover, the residuals of both regressions are stationary (with test statistics –5.21 (*p*-value 0.00) and –4.97 (*p*-value 0.00), respectively, each being significant at the 1 per cent level). The variables are cointegrated so that there is no sign of spurious correlation.

But indeed, data selection and results in Table 4.7 yield a strong argument in favour of speculation mainly originating in the futures market rather than in the spot market. The observation that a high oil price is associated with lower capacity utilization is made with US data. We argued that oil production outside of OPEC broadly follows the logic of a competitive market. It is therefore difficult or close to impossible for a single oil producer to manipulate the price of oil by accumulation of stocks. Competition punishes such a company through lower profits. This has

Table 4.7   *Correlation between the oil price and capacity utilization and oil inventories, 2000–14*

	(4.51)		(4.51')	
constant	12.63***	(2.95)	5.70**	(2.88)
capacity	0.03	(0.07)		
utilization	−1.21***	(0.29)		
spare			0.05***	(0.02)
Δin	5.48E−05***	(1.95E−05)	5.64E−05***	(1.99E−05)
ip	2.50***	(0.40)	2.56***	(0.37)
oilprod	−0.45	(0.66)	−0.30	(0.66)
exch	−2.68***	(0.26)	−2.55***	(0.28)
R^2	0.95		0.95	
DW statistic	0.54		0.50	

Notes: Standard errors are in parentheses. Coefficients with *, ** or *** are significant at the 10%, 5% or 1% level.

Source: Author's elaboration.

been discussed in detail before. As a consequence, in light of our empirical results, it becomes even more likely that speculation affects the oil price through the futures market.

By combining this outcome with the preceding investigation of the connection between monetary policy and the oil price and with our SFC model, we get an economic interpretation of it. Without being able to assess every detail, test results are broadly in line with our theoretical analysis. We notably find confirmation for the argument that monetary policy has an effect on speculation, which itself influences the price of crude oil. To sum up, policy transmission through fundamentals is insignificant and thereby indeed corresponding to its ambiguous character as argued in our theoretical analysis. Transmission through financial markets is more clearcut and seems to actually take place.

4.5   FROM OIL PRICE TO OIL QUANTITIES

As announced from the beginning, we are not only interested in the oil price but also in oil quantities. The effect of monetary policy on crude oil production and consumption can be seen as a two-stage transmission. Price variables in the oil market are more flexible than quantity variables, so that the first stage is the effect of a change in the interest rate level on the oil price. The second stage consists of the impact of an altered oil price on oil production and consumption.

In the theoretical part we argued that a high oil price triggers real investment in the oil industry, which pulls the oil price down again. In the presence of speculation, fundamental forces are distorted. There is a contradiction between price-raising financial investment in the futures market and price-lowering real investment in the spot market. Sooner or later, high production capacities and possibly high oil inventories lead to market pressure lowering the oil price. The more supply in the spot market increases, the more speculators become aware of changing fundamentals. Consequently, they stop betting on a rising oil price. More and more of them even expect a falling price and thus raise short positions. The price effectively starts declining. In comparison to the beginning, production capacities are higher. When the intensity of speculation has come down to its initial level, the oil price is therefore lower than initially. Crude oil is now available in abundance, which tends to raise consumption and thus oil intensity of total economic output.

To measure the effect on quantities, it would again be wrong to take oil production as the relevant variable. As explained above, it accommodates oil demand as long as exhaustion of reserves is not a binding constraint. What we need to investigate are the underlying forces on the supply side that determine the volume of oil supplied at a specific oil price level. For example, these are production capacities. In our case, capacities are reflected by the index of drilling oil and gas wells introduced above. The data series is integrated of order zero at the 5 per cent level and of order one at the 1 per cent level, respectively, as revealed by a corresponding ADF test. It cannot be wrong to take it in first differences. The advantage of this index is its short-term flexibility and, hence, its ability to react to changes in the short run. We may instead use, say, a variable that covers total capacity in the crude oil market by measuring total US oil output or potential output (global data at useful frequency are in any case hard to find) but it would be potentially biased severely by middle- or long-run developments like changes in the US share in total global oil production or technological progress. Our interest concentrates on the reaction of production capacities to changes in the oil price rather than other potential sources of change. The index has the outstanding feature that it measures well drilling as a particular economic activity, that is, the effort of investors who are devoted to increasing production capacities. It is therefore suggested as being least biased by other influences, since it does not measure final productive potential but, rather, the newly allocated resources in reaction to the price signal. We will have a look at final installed production capacities below.

The investigation of crude oil inventories showed that a higher oil price is not associated with significantly higher production capacities. It would nevertheless be a mistake to conclude that the oil price and production capacities are independent of one another. Correlation implies simultaneous effects and does not account for time lags. However, production capacities require time to be realized. Table 4.8 exhibits that extraction capacities increase in response to a higher oil price. The test results are highly significant at the 1 per cent level. This shows that we are able to find empirical evidence of a causal chain from monetary policy to oil market quantities.

A counterargument may now claim that investment triggered by an increasing oil price just takes place to the extent that it is required by an accelerating oil demand. It might be growing industrial production that raises the oil price in order to give an allocation signal to the supply side. Investment then takes the oil price back to its initial level. Investment is nothing extraordinary and therefore does not have an effect on the oil intensity of the economy. To assess this argument, extraction capacities are put in proportion to global industrial production. We call this ratio *cap/ip*. This series is stationary. Yet, it is far from being stable. This may be due to never-ending feedbacks between the oil price and drilling activity: from a neoclassical or, rather, a new Keynesian perspective, the market does not come to equilibrium, because feedback mechanisms face lags and hence are at work in every moment in order to adapt to past changes. Alternatively or additionally, one may suggest that the oil price contains a speculative component. As a consequence, real investment in the oil industry also reacts to other effects than just a change in spot oil demand. This idea is confirmed by econometrics, the results of which are in the second row of Table 4.8. There is significant Granger causality at least at the 5 per cent level but mostly at the 1 per cent level from the oil price to the *cap/ip* ratio. Even though this is not a proof, it enlarges the potential of the financial market effect further. At the very least, there are significantly more interrelations between oil price and quantities than fundamentals would suggest.

Table 4.8   Granger causality from oil price to oil and gas well drilling, 2000–14

Granger causality test from a to b	Significance	Sum of coefficients in line with theory	Causal relationship
oilpr → capacity	yes	yes	strong
oilpr → cap/ip	yes	yes	strong

Source: Author's elaboration.

Notice that a lower value of drilling activity does not mean an actual decrease in total production capacities. Such a decline is only given if, in a given moment, more exhausted wells are closed than new ones are drilled. Total installed extraction capacities therefore depend on many factors, particularly on the volumes of reserves. Reasonably easily accessible reserves are opened up first, so that each additional well tends to provide less oil. On the other hand, higher price levels make new reserves profitable and thus even accessible. And finally, technological changes affect total production capacities as well. The magnitude of total production capacities reacts with more delay to changes in other variables than drilling activity. As installed capacities consist of heavy industry equipment, they probably do not respond to short-run disturbances in the crude oil market.[9] Econometric analysis thus becomes nearly impossible, and it is more promising to rely on qualitative arguments. Effectively installed US oil and gas extraction capacities from the Board of Governors of the Federal Reserve System (2015b) are plotted in Figure 4.10. The dataset is seasonally adjusted but its relative smoothness would probably also remain if there was no adjustment. Obviously, extraction capacities strongly increase from 2006 onwards. Accounting for time lags of capacity construction, this

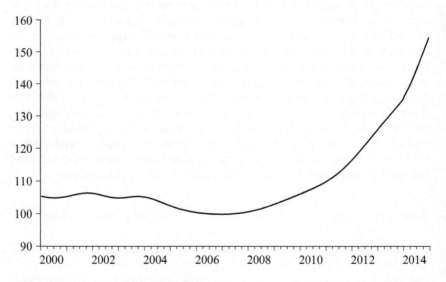

Source: Board of Governors of the Federal Reserve System (US) (2015b). Industrial Production: Mining: Drilling oil and gas wells.

Figure 4.10 Total installed US oil and gas extraction capacities, 2000–14 (2007 = 100)

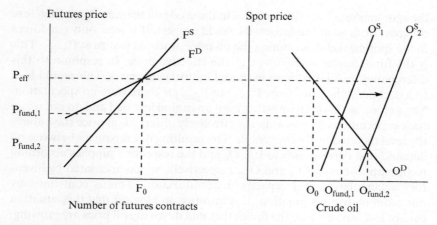

Figure 4.11   Real oil industry investment and the connection between the spot and futures market

Source: Author's elaboration.

coincides well with the increasing oil price and its persistently high level until mid-2014. The sharp price drop in the course of the financial crisis is not mirrored. This may be due to the inflexibility of existing extraction plants and to the fracking boom since about 2008. Figure 4.10 is broadly in line with the hitherto findings and the argument that the extended period of the high oil price has raised oil industry drilling activity, so that we end up with impressively large extraction capacities installed.

To conclude the argument, it should be tested whether additional investment and production capacities pull the oil price down from its high level, which we suggest to be determined by financial investment in the futures market. There is a difficulty in measuring an effect of the supply-side variable on the oil price. We assumed from the beginning that the oil price contains a fundamentals component and a financial market component. Even though these two components can hardly be separated in reality, we had to find a way to deal with them. The fundamentals component has been estimated separately. The financial market component is the residual.

At this point, we must come back to the dual nature of crude oil as a commodity and a financial asset. Let us show this in Figure 4.11, which is closely related to Figure 2.1 discussed in abundance in our theoretical analysis. We leave the slopes of supply and demand curves unchallenged for now, because any reasonable modification would not affect the argument. Assume the situation where financial investment in the crude oil futures market has driven up the futures price, which directly translates to the spot market. The effective price in the crude oil market, P_{eff}, is therefore different from what fundamentals would suggest. If it were only real forces in the spot market determining the oil price, it would realize at $P_{fund,1}$. This is the fundamental component of the effective price. In response to this discrepancy, real investment in the oil industry increases. This would lead to a decline in the price from $P_{fund,1}$ to $P_{fund,2}$ if there were no speculation. Yet, as long as financial investors keep up capital invested and do not revise their expectations downwards, the effectively observed oil price stays above the level dictated by fundamentals. The resulting discrepancies between oil demand that corresponds to P_{eff}, O_0 and the respective supplies according to the supply curves O^S_1 and O^S_2, respectively, are compensated by inventory accumulation and capacity underutilization. This is confirmed by our econometric examination. It is intuitive and logical that this situation cannot last forever since the forces that pull down the oil price are growing. Sooner or later, financial investors change their expectations and move more and more to short positions. This will bring the oil price definitely down. But uncertainty tells us that there is neither a rule nor a stable relationship that would make it possible to forecast when this turnaround in the futures market happens.

This raises the econometric problem that the price effect of real oil industry investment cannot be measured as long as speculation keeps the oil price at the elevated level. In contrast, there should be a direct and measurable effect on the fundamentals component of the oil price, provided that the fundamentals component is itself measurable. For this reason, the imperfect but hitherto useful calculated fundamentals price is employed again.

Consequently, we test for a causal effect from production capacities or drilling activity to the fundamentals component of the crude oil price. Whether calculated with net long positions or total positions, the price component is I(1), so that we employ it in first differences.

Granger causality tests, the results of which are depicted in Table 4.9, clearly reveal that larger production capacities lower the fundamentals component of the oil price. Once again, the argument may arise that this conventional feedback effect from production capacities to a price variable is only the expression of evolving fundamentals. However, the second row of Table 4.9 confirms the causal effect as a larger *cap/ip* ratio also reduces the fundamentals component of the oil price.

Given that there is a measurable effect on the fundamentals component of the oil price, we can conclude that there is also an effect on the whole price that cannot be estimated but is nevertheless present. The financial market effect produces a lot of noise and may go in the opposite direction to the fundamental component in a specific moment. Nevertheless, the

Table 4.9   Granger causality from oil and gas well drilling to the fundamentals component of the oil price, 2000–14

Granger causality test from a to b	Significance	Sum of coefficients in line with theory	Causal relationship
$capacity \rightarrow p_{fund}$	yes	yes	strong
$cap/ip \rightarrow p_{fund}$	yes	yes	strong

Source: Author's elaboration.

negative impact of real oil industry investment on the oil price exists *ceteris paribus*. The price effect through fundamentals affects futures market investment. Relaxed conditions in the spot market lower the attractiveness of crude oil futures for financial investors. Therefore, when the fundamental component of the oil price declines, the speculative component will sooner or later decrease as well. It is uncertainty, specifically in financial markets, which does not allow us to have a definite rule of when this happens. The relationship between production capacities and the effective price of crude oil thus becomes non-linear.

4.6   CONCLUDING REMARKS ON THE EMPIRICAL ANALYSIS

Our empirical analysis has confirmed our theoretical intuitions. SVARs have proven to be useful but limited in their ability to model the connection between monetary policy and the crude oil market, in particular the theoretically suggested financial market effect. This has led us to an alternative approach starting with a cointegrating equation of the oil market. The impact of monetary policy has in detail been investigated by testing for causal relationships between relevant variables. The finding that any single variable seems to be unable to represent the phenomenon of speculation in the futures market required a simplification in favour of measurability. We represented the financial market effect as a residual not explained by market fundamentals. This might be criticized as being rather arbitrary. But the fact that there is a significantly negative effect from the US federal funds rate of interest on this residual coincides exactly with our theoretical prediction, so that the suggested financial market effect is likely to be more than an arbitrary residual. The same effect could not be shown with significance in the period of unconventional monetary policy. A great part of this insignificant result may be due to the problematic issue of measurability of quantitative easing effects.

As a confirmation, we employed an alternative approach to crude oil inventories by the introduction of fluctuations in capacity utilization in the oil industry. There is strong evidence that inventories accumulate and capacity utilization decreases in times of a high oil price. This is obviously a sign that the crude oil market is not only driven by real forces but also by speculation, to wit, there exists a financial market effect.

A higher oil price gives an incentive for real investment in the oil industry. Our econometric test strongly supports this view. Conversely, higher production capacities lower the oil price. As a result, we end up with a conclusion in line with what has been suggested in our theoretical analysis and shown by our SFC model: an expansive monetary policy transmits ambiguously, if significantly, through fundamentals (besides the unambiguous price effect of a changing exchange rate) but significantly through financial markets. Financial investment in the futures market increases and drives the oil price up when interest rates are low. This triggers real oil industry investment, which raises production capacities and lowers the oil price again. Finally, oil is produced in abundance, thereby lowering the oil price. This is just the situation at the end of 2014: total extraction capacities installed reached previously unseen levels, as shown with the example of US data in Figure 4.10, and oil price fell by almost 50 per cent within half a year (see Figure 1.3). Given that the price elasticity of oil demand is larger than zero, this must raise oil consumption and the oil intensity of total economic output.

One may be fully confident in perfect competition and argue that the resulting lower oil price will lead to losses for oil producers and consequently again reduce overcapacities. The character of crude oil as a natural and exhaustible resource implies that oil reserves are accessible at a different production cost. A higher oil price makes new sources profitable. But likewise, a falling price makes them uninteresting. This is, of course, not completely wrong. But, first, during which time span is this going to happen? Speculation is a phenomenon that reacts immediately to new events and so does the oil price. Oil market quantities, however, do not adapt that fast to a changed environment. Extraction of crude oil requires high fixed investment. Once these expenditures are made, they count as a sunk cost, meaning that they are broadly irrelevant for daily production. In this situation, the difference between the oil price and variable production cost determines whether the extraction of an additional unit of crude oil is profitable. Hence, new investment in the face of a low oil price is not lucrative, since investment expenditures are too high. For existing production industries, however, such as those realized during the preceding oil price boom, oil production continues to be financially interesting, because fixed investment is a sunk cost. Furthermore, they may desperately need to raise oil output, because each barrel contributes to the financing of past investment expenditures as long as the oil price exceeds its variable production cost. The oil price may therefore remain low while oil production continues to be elevated for a considerable time.

Such a situation can trigger additional effects. Contrary to the expectations of many observers, OPEC has only slightly cut oil production to counteract the sharp fall in the oil price from mid-2014 (see, for instance, Krauss & Reed, 2015; Reed, 2014, 2015). One may speculate about the reasons for this decision. One of them might be that the cartel wants to keep its global market share. In this case, its intention is to keep the oil price sufficiently low for a sufficient time to beat competitors whose access to reserves is less privileged, so that they have to produce at a higher cost. OPEC probably eyes the newly installed wells during the preceding price boom. The resulting overcapacities caused by price-driving speculation then are likely to be strengthened over the medium run.

NOTES

1. For the stock and flow matrices as well as an overview of the model, see Appendix I.
2. Some of the following ideas are taken from my term paper 'Monetary Policy, Oil Price and Speculation' developed in a time series analysis course at the Summer School of the University of St Gallen, Switzerland, in June 2014.
3. For details about information criteria, see, for example, Enders (2014, pp. 69–70).
4. We regard the federal funds rate as exogenous, as argued by the theory of endogenous money. The partially endogenous character arises from central bank reactions triggered by economic developments.
5. For reasons of lack of space, we renounce additional steps to test for heteroscedasticity. In view of the quite strong significance of the variables estimated after removing autocorrelation, we suggest that the relationship is sufficiently stable and should not be greatly reduced by further corrections.
6. Here and in the following, ADF test results are not shown in separate tables in order to save space.
7. There are views denying any significant effect of monetary policy on output and thus the use of the interest rate to govern wealth and income distribution instead of output. For a review of approaches, see Rochon and Setterfield (2011).
8. The pattern can be explained by the behaviour of oil producers in the model: the shock to the precautionary demand parameter lowers inventories immediately and raises producers' profits, giving rise to higher capacity utilization and overshooting investment expenditures. Resulting excess production capacities lead to renewed accumulation of inventories. The second shock of precautionary demand back to the initial parameter value makes inventories jump, since production is adjusted only with lags. From the new point, disinvestment and reduced capacity utilization reduce production, and hence inventories, step by step.
9. For instance, while drilling activity may react fast to a price decline, installed capacities are not shut down immediately.

PART III

Achieving stability and sustainability: economic policy making

5. Economic policy propositions: an overview

In this final part, we will analyze the problems resulting from the relationships between monetary policy and crude oil price and quantities. Thereafter, we discuss a set of possible policy propositions. It will be crucial to make use of the hitherto developed insights into the working of monetary policy and the crude oil market.

5.1 ECONOMIC AND ECOLOGICAL PROBLEMS ARISING FROM THE CRUDE OIL MARKET

We have shown that, through speculation, monetary policy cannot only affect the crude oil price but also crude oil production and consumption. They are a result of the logic surrounding how the futures and spot markets are integrated. Against this background, two problems arise, of which one is an ecological and the other an economic nature. The former has already been outlined although without having been denoted as a problem: a higher oil price induced by speculation raises real investment in the oil industry, which itself enhances oil supply. The price falls to an even lower level than before with the corresponding positive response on the demand side. Oil consumption tends to increase so that the oil intensity of economic output (or the oil intensity of society's consumption in general) is higher than it would be without speculation. We end up with higher pollution and corresponding contamination of the world climate (see, for example, Murray & King, 2012). In the long run, the development of alternative energy sources and improvement of energy efficiency is weakened.

The second problem concerns the issue of both financial and economic stability. While both terms are broad concepts, the former is particularly difficult to define (for an extended debate about defining financial stability, see Panzera, 2015, pp. 8–23). Schinasi (2004, p. 8) chooses the following definition:

> A financial system is in a range of stability whenever it is capable of facilitating (rather than impeding) the performance of an economy, and of dissipating

financial imbalances that arise endogenously or as a result of significant adverse
and unanticipated events.

Our interest is not in the whole financial system but rather in the crude oil
futures market. To the extent that oil price speculation exceeds the hedging
needs of oil producers and consumers, it loses its beneficial impact. By
moving the oil price away from what is suggested by the spot market, the
futures market starts impeding rather than facilitating the performance
of an economy. According to our analysis, price fluctuations caused by
financial investment give rise to imbalances in the spot market. This is
where financial instability becomes economic instability. Chant (2003,
p. 4) clearly distinguishes between financial and macroeconomic instability by characterizing the latter as caused by aggregate demand and supply
shocks, such as changes in expenditures or technological progress. The
term 'macroeconomic' requires closer consideration in the context at hand.
The common debate about financial stability concerns the economy as a
whole (see Panzera, 2015, pp. 8–23). If an event affects the economy as a
whole, it is of a macroeconomic nature; if there is only a change within
some parts of the economy without changing the total economy, it is a
microeconomic one (Cencini, 2012, pp. 54–55). Testing our case against
this strong proposition, we find that even though the crude oil market is
only a fraction of the economy, it has effects on the whole economy, as we
will argue. In accordance with our preceding macroeconomic analysis, the
investigation of policy propositions continues to be macroeconomic. For
convenience, when we are not talking about stability in the whole economy
but about stability in either the crude oil market or its opposite, the non-oil
economy, we will talk about 'economic' stability hereafter.

The instability problem consists of the economic risks of both real
and financial overinvestment. On the one hand, speculation may create a
bubble in the futures market by moving the oil price away from what fundamentals would suggest it to be. This bears great risks. Financial investors
who bet on the long side of the market incur large losses once the price falls
again. Private bankruptcies have their corresponding repercussions on the
banking system of an economy. Their dimension may be hard to evaluate
owing to crude oil futures often being only a fraction of an investment
portfolio. On the other hand, the response of oil companies to a high oil
price by raising real investment potentially leads to the same bubble in the
crude oil spot market. As we pointed out, the price therefore has to fall
inevitably. The conventional feedback mechanism would imply that the
oil supply would decrease again. We explained earlier why this feedback
may not take place so soon and that in the medium term, there may even
be opposite effects of rising supply. Yet, if oil production is not profitable,

it will not persist in the longer run. According to production costs, different oil sources require a different minimum oil price level for sales to cover those costs. The IEA (2008, pp. 217–219) calculates the production costs of conventional crude oil to be between less than 10 and 40 US dollars while, for instance, those of oil shales range between 50 and more than 110 US dollars depending on the specific source. Those producers, whose costs are below the price or just move around it, will shut production down sooner or later. Going bankrupt, however, is a process that leaves its traces in the form of unpaid credits and suspension of employees. Since in the oil industry all companies depend on the same global price level, a drop of the latter may lead to a crisis in a whole branch of economic activity. In the case of the United States, for example, there are more than 400 000 stripper wells all over the country contributing 11 per cent to US crude oil output (Meyer, 2014). This shows that a decrease in the oil price can have far-reaching effects in the sense that a great number of entities are concerned. Besides crude oil, other energy sources like natural gas are likely to be concerned as well by the high oil price, owing to the cointegrating relationship we suggested above. As soon as enthusiasm about the fracking boom entered the stage when the oil price was again around 100 US dollars per barrel in 2009, critical voices rose as well, stating that production returns were not sufficient to cover investors' debt (see Ahmed, 2013). After the oil price had dropped in mid-2014, troubles of oil-producing regions and companies appeared in the daily press (see, for instance, Scheyder, 2015).

Finally, fluctuations in the oil price affect the rest of the economy where fossil fuels are an important production input. We already mentioned different investigations in this respect (see, for example, Bernanke et al., 1997; Blanchard & Galí, 2007; Kilian, 2010a, 2010b). This is the point where the stability issue definitely becomes macroeconomic. Financial and macroeconomic stability are usually separated in terms (see Chant, 2003, p. 4). There are, naturally, connections in between. An oil price change due to financial investment in the futures market directly affects the real economy. Owing to the dual nature of crude oil as a physical commodity and as a financial asset, financial and macroeconomic (in)stability are the two sides of the same issue. Importantly to note, financial stability includes the ability of the financial system to facilitate the efficient allocation of resources (Schinasi, 2005, p. 2). Hence, not all changes in the oil price are a sign of instability. We focus on price changes originating in financial speculation that do not correspond to any need of the oil spot market.

We argue that monetary policy may be a possible cause of such over-investment in the crude oil market. A similar line of argument is provided by White (2013, pp. 33–36), who denotes it as 'malinvestment' caused by easy monetary policy: low interest rates make investment credit cheap and hence may lead to investment expenditures that can no longer be afforded when interest rates rise again. However, there are differences to this view. Mainly, we do not adopt the concept of the natural interest rate nor do we connect our findings to it in any way (ibid., p. 30). Its application in a Wicksellian sense means that a deviation of the financial interest rate from the natural one inevitably leads to a changing price level and mal-investment as we have already outlined in the chapter on monetary theory. Our conclusion does not share this view. On the one hand, the natural interest rate cannot be observed. All conjectures about its level are there-fore speculative (Rochon, 2004, p. 16). On the other hand, macroeconomi-cally, expansive monetary policy may contribute to growth in total output without creating imbalances in contrast to what neoclassical theory sug-gests (see, for instance, Colander, 1996, pp. 28–29; Sawyer, 2002b). There is nothing of a harmful disequilibrium herein *a priori*. With respect to the market for crude oil, we do not state that monetary policy inevitably gives rise to overinvestment by deviating from the natural rate of interest. The effects detected in our analysis are due to the possibility of speculation and the dual nature of crude oil as a physical commodity and a financial asset. These two features cause a reaction of the crude oil market price and quantity variables to monetary policy that is stronger than in the rest of the economy.

Both the ecological and the economic problem arising from our analysis require a policy response. Neoclassical economists may recommend the removal of market rigidities in order to enable the economy to smoothly follow its equilibrium path. Such ideas are motivated by the argument that, in past decades, decreased rigidities have mitigated the impacts of oil price shocks on the economy (see, for example, Blanchard & Galí, 2007). Yet, first, our results are not founded on rigidities of a new Keynesian type. They will always exist in capitalist economies, because they are a precondi-tion of capitalist production (Cottrell, 1994, p. 591). Hence, one cannot say that a competitive market, as it exists in the real world, can overcome the effects of monetary policy and speculation on the natural environment or on economic risks.

If economic policy is to intervene in the crude oil market – be it by government activity, new regulation laws, an improvement of the financial structure of the economy, or a change in the conduct of monetary policy – it has to be aware of the desirable outcome. Consequently, we first have to determine the optimal level of the crude oil price.

5.2 WHICH LEVEL OF THE CRUDE OIL PRICE IS OPTIMAL?

In general, one might take the view that prices do not have a normative aspect, since they are determined by market forces and are thereby, so to speak, a natural product. In a similar, more sophisticated way, it can be argued that speculative price effects are harmful but that the fundamental component of the price represents the 'true' price. However, as argued above, it is hardly possible to distinguish price components. Alternatively, commentators take the perspective of the supply or demand side. This can be shown with regard to food commodity prices. There was an analogous debate in recent years, similar to the one about oil prices (see Baek & Koo, 2014). Most voices raised great concern about the consequences of high food prices for the poor of developing countries and, especially, the urban population. However, high food prices help producers to raise their income, which is to the benefit of hundreds of millions of farmers around the world. It becomes clear that debates about the optimal price cannot avoid political considerations and motivations.

Even though this concept is still rather abstract, it provides some point of reference. With oil, however, the case is much less clear. One may again come to the same conclusion by only taking producer and consumer welfare into account. But there is an additional aspect specific to crude oil. All our arguments and, specifically, the SFC model, go in the direction that, despite non-linearity, a high oil price *ceteris paribus* goes along with lower oil consumption relative to a low oil price. Thus, the oil price is relevant for the amount of energy use and pollution. A high oil price may be seen as advantageous, since it gives both producing industries and individual consumers an incentive to raise energy efficiency of buildings and production processes. Moreover, they substitute other, potentially more sustainable and renewable, energy sources for oil. However, this is only the demand-side effect of the oil price. The supply side is more often neglected. The higher the oil price, the more oil reserves – also those that are not easy to access – can be developed and extracted. Oil sands and extra-heavy oil face higher production costs than conventional oil (IEA, 2008, pp. 215–216). Moreover, their extraction requires technologies that produce chemical residues, which run the risk of being distributed in the natural environment, especially in freshwater systems (see, for example, Hodson, 2013). The environment, including climate, benefits from a high oil price with regard to the demand side of the oil market but is damaged with respect to the supply side.

Against this background, it is not easy to characterize a price. Is it a moment's snapshot describing the state of the market? Or is it a dynamic issue reflecting the underlying dynamics that lead to this price as well as the dynamics the price will trigger? In the case of the latter, a price cannot be understood in a single moment, only in a period of time. From this point of view, not the price level is relevant but rather its change. In fact, this is the only way we can get an idea of what a price really means. A high price cannot be judged as such, because it bears the dynamics in it that will lower it again (if we abstract from the exhaustion of oil reserves, in our case). With the presence of speculation, these dynamics get an even larger magnitude. This aspect is especially important with respect to the problem of overinvestment in the oil industry in the course of price speculation. The level of the oil price then is of secondary priority. The crucial issue is the extent of price changes. Stable investment conditions require a stable price.

To achieve the ecological goal at the same time, we have nevertheless to find a level of the price that goes along with sustainable levels of oil quantities, that is, production and consumption. In the face of climate change and environmental contamination, it is desirable that the quantities are as small as possible. In contrast, assuming away the ecological challenge and merely focusing on market stability, oil quantities become basically irrelevant. The key variable for stability is the oil price. Given that it is stable, stability of the crude oil market as a whole is guaranteed whether oil production and consumption are large or small. These reflections are the basic preliminary for the development of economic policy propositions that are able to address the problems resulting from monetary policy and futures market speculation.

5.3   SEVERAL OPPORTUNITIES FOR ECONOMIC POLICY

There are several ways to encounter the problems identified above. We will debate and criticize them according to suggested advantages and drawbacks. Some are better known than others and they rely on different approaches. We will also develop drafts of alternative policy proposals. While some will require smaller steps from the present situation, others are more ambitious. Especially, and this may be new in this debate, monetary policy will be put to the fore. Overall in this work, therefore, it is not only the role that monetary policy plays today with respect to the crude oil market that is taken into account. We go further and ask also about the role monetary policy may potentially play in an economic framework that intends to overcome the problems of instability and the ecological challenge associated with crude oil.

5.3.1 Enhancing Stability in the Crude Oil Market

We start by discussing policy proposals for the issue of economic and financial stability in the crude oil market. The ecological problem will follow thereafter. With each of the policy approaches, we will discuss its aspects, and if and how it can be brought in line with stability measures previously analyzed.

Financial market regulation

When speculation occurs, the call for regulation follows on step. In commodity markets, the conception aims at making speculative trades impossible, so that futures markets only serve the needs of commercial traders, that is, producers and consumers, to hedge their purchases and sales, respectively. The proposed measures are most articulated with regard to food commodities and have got larger public attention since 2008, when food prices climbed and revolts in numerous developing countries erupted. The call for regulation is in great part shared and spread by the altermondialist movement (see, for instance, Global Justice Now, 2015). Since all important commodity futures markets basically work in the same way, the findings and conclusions can be employed to the market for crude oil, too.

There are a couple of central measures in discussion. Some are integrated fully or partially in the US Dodd–Frank Act. In order to turn to each of the debated propositions in a structured way, we follow Staritz and Küblböck (2013, pp. 16–23), who separate them into six groups. Where it is not specified further, we replicate their arguments as they represent the overall debate of commodity market regulation quite well:

- *Transparency and reporting:* transparency in futures markets can be seen as the first step towards enabling any regulation at all. Reporting is required to have data of different trader groups and strategies. Division in commercials and non-commercials or even dividing the latter again in swap dealers and money managers (see CFTC, 2015) is not sufficient. As, for instance, Büyükşahin and Robe (2011) show, empirical results may differ when data disaggregation is improved. Nobody should be excluded from reporting and it should be guaranteed by national or international authorities rather than single exchanges.
- *Limit/prevent OTC trade:* OTC trade is often said to have become especially meaningful after the CFMA entered into force at the end of 2000, since it removed all OTC activity from regulation under the Commodity Exchange Act (CEA) (see, for example, Greenberger, 2010, pp. 99–100). Market risk therefore tends to increase in many

aspects. First, nobody knows the total exposure in the market. Second, capitalization of traders may be insufficient. In contrast to official exchanges, there are no definite margin requirements so that trades become even more leveraged. Regulation therefore claims that standardized contracts be only traded on and cleared by transparent exchanges that are themselves also well capitalized (ibid., p. 112). Outside of the exchanges, only non-standardized contracts should be allowed and they should also be subject to margin rules and general regulation requirements such as reporting. Moreover, the authorities should work in favour of far-reaching contract standardization in order to concentrate futures trading in a smaller number of exchanges and to minimize OTC trade (Staritz & Küblböck, 2013, pp. 19–20).

- *Position limits:* futures market participants are not allowed to hold a larger amount of contracts than a specific threshold. Such position limits reduce speculative activity. Futures market transactions then should merely be of an amount required to guarantee liquidity for price discovery and to satisfy hedging needs of commercials. Thereby, gross as well as net positions should be taken into account. This claim is supported by our suggestion that the importance of speculation cannot be captured by a single variable. In the same way, position limits should be applied to individual traders as well as to aggregate specific trader classes. There should be no general exemptions for any trader class, not even commercials. Exemptions should only be granted to specific transactions insofar as they serve hedging necessities.

- *Price stabilization instruments:* in most stock exchanges, there exist regulating tools leading to a standstill if prices fluctuate too much within too short a time span. Yet, there is nothing of the sort in commodity futures markets. Speculative trading is therefore free to move futures prices. The regulating authority may use a financial transaction tax between 0.001 per cent and 0.1 per cent as a price stabilization instrument. It thereby defines a price band within which the commodity price is suggested to lie according to fundamental conditions. Once the price leaves the price band, the transaction tax becomes effective. Each transaction in the futures market then is taxed by a given rate, which might vary depending on the degree of price deviation. Financial investors – either on the long or on the short side of the market – who aim at benefitting from very small price changes then lose their profit prospects. They lower the volume of transactions and hence reduce the impact on the price. The latter thereby moves back to the predefined price band.

- *Restriction of certain groups of traders or trading strategies:* high-frequency trade is a strategy where traders hold financial assets only for milliseconds before selling them again in order to benefit from infinitesimal price changes. Another and closely related problematic issue in this framework is algorithmic or technical trading. Speculators use technical systems to calculate past price trends and to trigger an automatic purchase of long or short positions depending on the trend these systems perceive (Schulmeister, 2012, pp. 38–40). Such systems may give rise to uncontrolled price fluctuations. Some of these strategies might already be counteracted by a financial transaction tax. Otherwise, they can be restricted or prohibited.
- *International cooperation and supervision:* finally, for regulation to be definitely effective, it should be extended globally in accordance with global commodity markets. International cooperation should be strengthened to harmonize national regulations or to set global minimum standards. In the optimum, there is a global authority with regulatory and supervisory competences.

An advantage of the proposed measures is that they can be directly applied to the place where speculation occurs, that is, in the crude oil futures market. They do not require a full macroeconomic framework to enter into force. Basically, the claims listed here are rather a question of legislation than of economic reasoning once their general idea is given. However, the seeming practicability of the regulation approach is as well a drawback. Besides well-known doubts, it leaves some important economic issues unquestioned.

The first objection that occurs regularly is the one of legislative effort, which produces a lot of bureaucracy. Especially with regard to the regulation or, rather, strong reduction of OTC trade, it requires a lot of supervision without a guarantee to be fully successful. Moreover, regulations tend to leave gaps used by financial investors to circumvent the rules in a legal way. For instance, if financial investors face position limits, they may contract on a physical OTC delivery and hence can claim in the futures market that they have a commercial interest and get rid of the position limits (*The Economist*, 2013). Regulation may be more or less efficient and effective and thus leave more or less regulatory gaps open.

The regulation approach does not ask the question about the oil price level that we sketched out above. Therefore, it does not seem to have an idea of the preferred price, whether it should be relatively high or low. The exclusive purpose is to remove speculation from the futures market. The implicit assumption lying therein is the acceptance of the market price driven by fundamental forces to be the 'true' price. In order to ensure economic stability and to prevent bubble-building in commodity markets, this view may be appropriate. For those prices of food commodities that tend to be driven up by futures market speculation such that food becomes unattainable for great parts of world population, the argument seems to be adequate as well. Yet, for crude oil, which has not only an economic but also an ecological importance, simple elimination of speculation will not be sufficient, as will be shown later.

By, directly or indirectly, identifying the fundamentals price as the 'true' one, the regulation approach suggests that the fundamental and the speculative components can be separated. In our theoretical analysis where we distinguished transmission channels of monetary policy through fundamentals as well as through financial markets, we argued that they are closely related and interact dynamically. Their numerical separation is therefore not possible. In the empirical part, a concession in this regard was made to test for basic significance of fundamental and financial market price effects. Yet, we are far from claiming to have found exact numbers attributed to each one.

For these reasons, it becomes quite hard to define an oil price band outside of which a financial transaction tax is to be imposed. Where should the limits be set such that all movements within the oil price band can be attributed to changes in fundamentals while those outside of it are purely speculative? Likewise, where are position limits for financial investors to be set, so that they serve to provide liquidity but do not exceed those needs? Uncertainty about these issues makes regulation in the form of conventional legislative frameworks rather inflexible. Since the speculative price component cannot be detected, the regulatory framework cannot be determined by means of objective economic arguments. It is rather subject to permanent and renewed economic analysis and policy. For regulation to be effective, permanent market supervision and possible adaptation of regulatory values are necessary. The supervising authority should therefore be endowed with a certain degree of flexibility and independence, as it is well established regarding monetary policy.

Counteracting oil price fluctuations
Instead of creating a legislative framework to rule out speculation, another approach consists in active intervention in the crude oil market. Whether such actions are executed by the government administration or by some specified regulatory authority is of secondary importance here. The basic idea consists of a reaction to price developments that impact the market in the opposite direction. This measure is strongly recommended by Davidson (2008). He mentions the US strategic petroleum reserve. While it fulfils the emergency stocks obligations of the IEA and is argued to serve

as a national defence fuel reserve, it mainly 'provides the President with a powerful response option should a disruption in commercial oil supplies threaten the U.S. economy' (Office of Fossil Energy, 2015). In March 2015, for instance, the strategic petroleum reserve stocks amounted to about 36 days of US petroleum consumption (EIA, 2015b, 2015d). In a moment when the oil price is very high and speculation is suggested to be a major price driver, a partial release of the strategic petroleum reserve by selling it on the spot market pulls the oil price down as supply conditions are relaxed.

It is an open question whether the oil reserves are sufficient to bring the price of oil to its desired level. According to Considine (2006) and Stevens (2014), the price impact of the use of the strategic petroleum reserve is quite modest, while Verleger (2003) argues it to be strong. On the one hand, one may argue that the public is aware of the strategic petroleum reserve and once it is released to its limits, speculative investment may carry on. On the other hand, financial investors often aim at benefitting from very small price changes, as in the case of high-frequency or technical trading. A one-time release of crude oil reserves may have a sufficiently large effect on the oil price, so that the losses incurred by speculators can no longer be compensated by small-scale profits. This might then be sufficient to free the oil market of the main speculative impacts.

The holding of strategic reserves is, however, suggested to be both insufficient and inefficient as it implies physical storing capacities and trading infrastructure. Nissanke (2011, pp. 54–56) and Nissanke and Kuleshov (2012, p. 35) propose market intervention through the futures market rather than directly in the spot market. In the case of a high oil price, the political authority should enter the oil market on the short side of it to lower the oil price. In this way, counteracting measures can be taken faster and more flexibly without physical exposure. Moreover, trading crude oil in paper form rather than as a physical object allows better fine-tuning. To be credible, the public trader has to convince financial investors to be ready to make losses if necessary to stabilize the oil price. The more credible the threat, the less intervention is actually needed, since speculators do not want to take too high a risk of investment losses.

This approach does not make the claim of being able to distinguish the fundamental and the speculative components of the oil price. In contrast to financial market regulation, it does not make suggestions about the origin of price movements to intervene directly at those origins after their identification. Rather, it lets market forces, in the spot market as well as in the futures market, act freely. It takes the final result, to wit, the crude oil price, as the starting point and then counteracts correspondingly. This way of proceeding is relatively simple and allows flexible and probably effective intervention. However, it also has its shortcomings. While it tries to stabilize the crude oil price, and hence stabilize overall market conditions, the stabilization effort may also have destabilizing effects. Owing to the indivisibility of the oil price in its fundamental and financial market component, market intervention cannot be precise in the sense that it manages to filter out exactly the speculative fraction of the price. It thus may be insufficient, which means that it only mitigates but does not fully break the speculative influence. Or it might be overshooting, implying that fundamentals are affected more than if only speculation was neutralized. Such an overshooting has effects on both supply and demand sides of the spot market by influencing the amount of inventories, production, consumption and investment behaviour. Such an outcome is more likely if counteraction takes place in the spot market instead of the futures market, owing to the greater flexibility of the latter. Concerning implementation, sophisticated international cooperation would be required for this approach to be successful. If each country pursued a different strategy of market intervention, effects would probably be more destabilizing than stabilizing.

5.3.2   Reducing the Share of Fossil Energy

The above-mentioned policy propositions in principle contribute to economic and financial stability. Given that they are able to ensure that the oil price corresponds to the underlying fundamental developments without containing any other component, a certain degree of ecological stability is also established. In the optimal case, the regulatory approaches and direct market intervention guarantee economic stability and prevent fluctuations of oil consumption that are harmful to the natural environment in general and to climate in particular. Otherwise, they may themselves be a source of smaller or larger disturbances and hence affect oil consumption volatility, too. Be that as it may, the instruments do not contribute to a clear improvement of the ecological balance of the economy but rather re-establish the *status quo*. Therefore, we may ask the question, how can the insights of the hitherto analysis be employed productively to get a step further?

So far, we established that monetary policy has an effect – mainly through the futures market by means of financial investment – on the oil price and hence on oil quantities. If climate change is to be stopped, production and consumption of fossil fuels and hence of crude oil must decline. There is hardly any serious doubt to this claim in science at the time of writing (see, for instance, Murray & King, 2012). The question of interest is, therefore, how economic policy can achieve not only economic stability and smoothing of oil consumption but also both economic stability and a reduction in oil consumption. We will develop a way that our analysis may serve these purposes, that is, how monetary policy and futures markets might be used in a helpful way. In the course of this proceeding, different propositions will be presented that are already frequently discussed in the sphere of economic and environmental policy and may already be implemented in related political fields. In the final chapter, we will see that our own proposition is a combination of already existing approaches that pursue the purposes of economic stability and ecological sustainability.

Oil supply target

Acknowledging that oil consumption necessarily has to decrease, stop or at least mitigate climate warming, the political authority may simply set targets for oil production. Similar propositions are made by various voices including economists, think tanks, and journalists (see, for instance, Barnes, 2008; Boyce & Riddle, 2007; the Foundation for the Economics of Sustainability (Feasta), 2008). The reason why it is again important to make a difference between production and supply will easily become clear.

The political authority may set a target either for oil consumption in absolute levels or for the oil intensity. In order to assess the oil intensity of the economy, ratios have to be defined. The most obvious ones are the proportion of oil consumption to industrial production or to total GDP. While the absolute level of oil consumption is the only relevant one from the view of climate policy, the intensity ratio takes long-run economic development and business cycles into account. The former features the problem of how setting a target without reconsidering the state of the economy can have negative repercussions on output. Using an oil intensity ratio is therefore more appropriate. The political dispute then consists of the level of the targeted ratio. A constant target allows for the stabilization of oil intensity but implies greater pollution if output grows. If oil consumption is to be decreased in absolute levels, the ratio must, therefore, fall over time, given that there is a long-run growth rate of output larger than zero. The condition for a falling oil intensity target to lead to a decreasing quantity of oil consumed is given by the following simple formula. The variable of interest, be it industrial production or GDP, is described by Y_t and is assumed to grow by an annual rate g_y.

$$g_y = \frac{Y_t - Y_{t-1}}{Y_{t-1}} \tag{5.1}$$

The oil intensity ratio R_t should fall enough so that crude oil consumption C_t is at least constant or otherwise declining, meaning that

$$C_t \le C_{t-1} \text{ which implies } C_t / C_{t-1} \le 1 \tag{5.2}$$

The percentage change of the intensity ratio g_R is therefore given by

$$g_R = \frac{R_t - R_{t-1}}{R_{t-1}} = \frac{R_t}{R_{t-1}} - 1 = \frac{C_t/Y_t}{C_{t-1}/Y_{t-1}} - 1 \qquad (5.3)$$

$$= \frac{C_t}{C_{t-1}} * \frac{Y_{t-1}}{Y_t} - 1 \leq \frac{Y_{t-1}}{Y_t} - 1 = \frac{1}{1 + g_Y} - 1$$

The condition is thus given by

$$g_R \leq \frac{-g_Y}{1 + g_Y} \qquad (5.4)$$

For small values of g_Y, the nominator can be roughly approximated by 1, saying that oil intensity has to decrease by the same rate at which output grows for oil consumption to remain constant. For the latter to decline, the drop of the ratio has to be larger than output growth. To account for year-to-year fluctuations in output, the target can be set for longer-period averages of, for instance, ten years.

Crude oil or petroleum consumption, respectively, takes place in a decentralized manner, that is, in industries and individual households. Political intervention being employed on the consumption side is thus quite complicated and runs the risk of ending up involving more bureaucratic effort than would in fact be necessary. The target hence may be set on the oil production side. Conditions in the oil industry are more or less competitive and thus more or less concentrated depending on the country. But even in the case of the United States, where there is also a large number of small producers present in the market, political intervention is still easier to carry into execution on the production side than on the consumption side (Feasta, 2008, p. 4).

Yet, production is not the same as consumption and hence the target of interest, that is, the oil intensity ratio, is distorted. Production includes inventory accumulation, which is *a priori* independent of effective consumption, especially when we take oil imports and exports into account (see below). It is thus more useful to take supply as the target value, to wit, the volume of crude oil that is effectively sold in the market. Supply is equal to demand in any given moment. Hence, the targeted ratio is now distorted only by a much smaller gap consisting of oil demand and effective oil consumption, which amounts to inventory building on the consumer side. As oil supply targeting is a medium- or long-run purpose and since it is reasonable to suggest that inventories of consumers follow a stationary pattern, the replacement of oil consumption by oil demand seems to be appropriate. Yet, the approach is called oil supply rather than oil demand targeting, because it is implemented on the supply side of the crude oil market.

Until now, this approach might seem rather abstract and radical to some readers. It is justified to ask how such a simple design of a proposal can be implemented in the real economy. With regard to the latter, however, it becomes clear that this proposal is not as unique as one might suppose. Indeed, its conception is related to greenhouse gas emission trading systems of which the largest one currently exists in the European Union (European Commission, 2015, p. 1). This system basically consists of a 'cap' that determines the maximum volume of emissions allowed in one year. The cap is composed of allowances that a company must possess by an amount that covers its annual carbon emissions. While the allowances were allotted to the sectors for free every year in the first two periods of the system's existence from 2005 until 2012, they have been allocated by an increasing share through auctioning since then (ibid., p. 4). Companies can trade the allowances among themselves in accordance with their emissions arising from production. The limitation of allowances by setting a cap gives them a price. The rationale behind the system is to let market forces work freely, suggesting that allowances are traded in a way that leads to an efficient allocation. If so, then, according to the suggestions, reduction of carbon emissions can be achieved in the most cost-effective way (ibid., p. 2). The system's basic economic framework can be traced back to Coase (1960), who states that, in the absence of transaction costs, all external effects can be internalized by agents' bargaining, once property rights are well defined. In this case, emission allowances are made a tradable asset over which companies bargain and thereby find the most efficient level of output and the optimal degree of contamination. A difference is, however, given by the fact that political action is required to provide emission with a price and to determine what the socially optimal, or acceptable, level of overall pollution is. Oil supply targeting could be implemented in an analogous way by emitting allowances according to the targeted level of the oil intensity. They are allotted to oil producers and define the total quantity of oil that is allowed to be sold.

Yet, the European carbon trading system is not at all free of criticism. As Gilbertson and Reyes (2009, pp. 9–12) put it, numerous reasons make carbon trading an inadequate and insufficient strategy to address climate change. For their first criticism, a cap set at too high a level, so that it exceeds effective current pollution of individual industries, does not put any pressure on the latter to improve production processes. Even more than that, they can sell excessive allowances to other polluters and thereby benefit financially from the emission trading system. Second, as far as the free allocation of allowances to economic sectors is concerned, historical patterns of industries' carbon emissions are considered. Those industries that face a relatively large volume of emissions are allotted with more allowances than others. This gives rise to a reward for industries that did not make any efforts in the past. Third, the UN-administered so-called 'Clean Development Mechanism' allows for emission reductions in developing countries that are not part of the trading system. When a company realizes a project in such a country, it gets emission allowances in the amount of carbon emissions saved by the project. The quantity saved is estimated by the difference between pollution of the actually implemented project and the project that would otherwise have been realized. However, such an estimate mainly consists of guesses as to what the alternative project would be. Those guesses may in principle be quite generous, so that the success of the system might appear impressive. Yet, what undeniably takes place besides artificial guesses about emission reductions is the increase in pollution as a result of the project realized. At the same time, the company in the north is allowed to augment emissions by the amount of the additional allowances. This offsetting opportunity consequently allows companies in the system to increase the number of allowances according to requirements (Bernier, 2007).

The approach proposed at this juncture should be able to circumvent these critical points. As regards the first one, the cap of permits (in our case permits to sell crude oil) has to be determined tight enough in order to have an effect on oil supply. In this respect, the rule derived above about how to set the target ratio becomes important. This is dependent on political decisions assessing the reduction goals to be reached. The second problem, that is, free allocation according to historical production, may indeed be considered as a reward of those who produced most in the past. Auctioning the allowances without any prerequisites would therefore be fairer.[1] Larger producers would have the power to purchase more permits but this corresponds to the distribution of market shares already in place. Offsetting emission reductions in countries outside of the trading system, which represents the third problematic point in the European emission trading system, should simply be kept out of the design. To get there, it is crucial to determine the working of the approach within a single country and its relation to crude oil exports and imports.

Implementing the oil supply target globally is the simplest and most effective way. However, in a world where global governance has not developed too far, this possibility is unlikely. In order for an oil supply trading system to be nevertheless employed, the latter must be compatible with the fact that most relevant legislation is national while the crude oil market is globalized. In a carbon trading system, the basic economic idea is to give

environmental contamination in general and carbon emissions in particular a price by making them scarce (European Commission, 2015, p. 2). By
setting a cap on total emissions, a market is created out of nothing. By
contrast, an oil supply target is set in a pre-existing global market. This
fact may provoke critical objections concerning the viability of such a
system in the market for crude oil. However, the setting of this approach
allows for quite an elegant realization. The problem would be acute if the
target concerned production instead of sales. In the former case, companies could produce no more than what is allowed by the cap. This would
leave the problem of determining the level of the target, since it is not clear
how oil exports should be weighted. The latter may grow in the future or
not be independent of the target set for national oil intensity. Constraining
oil exports would come at the cost of national producers while missing
exports would be compensated by other foreign producers. With respect
to crude oil imports, a production target can, of course, be easily circum-
vented by oil imports so that there is no incentive to use fossil fuels more
efficiently. Focusing the target on crude oil sales resolves the problem: oil
producers are allowed to produce in excess of what is allowed to sell in the
home economy and to export it. Oil importers, on the other hand, can in
principle import as much oil as they want, but they can only sell a quantity
that is backed by supply allowances. In this way, producers still have the
opportunity to participate in the global crude oil market as they would
without a supply target. On the other hand, the target is effective in the
national economy and can exert the desired effects on the oil intensity in
the country. Hence, there are no (legal) arbitrage opportunities that may be
used to circumvent the policy measure. The only potential repercussions
a supply target can have on the global market is the reduction in national
demand for oil. Yet, this effect is the very intention of this approach. To
the extent that it makes consumers employ more efficient and alternative
technologies, declining demand pulls the oil price back in the direction of
the world level.

This policy proposition cannot, however, get rid of all problems in
the theoretical sphere as well as in the area of practical implementation.
The tighter the supply target is set, the greater is the incentive of illegal
circumvention and hence the greater the effort of supervision required.

But there is a more serious problem from an analytical point of view.
While oil quantities are influenced by policy intervention, the oil price may
continue to fluctuate. Controlling both quantities and prices is a logical
impossibility in market economies. In a system with oil supply targeting,
the price of crude oil not only reflects intrinsic fundamental developments of the oil market but is naturally strongly influenced by the supply
target, too. To enable a smooth oil price development, the cap should be

set according to supply and demand conditions in the crude oil market. In the European carbon trading system, this was found to be quite a difficult task. In the first phase of the system's existence, the price per tonne of CO_2 started at a level of about 23 euros in 2005, climbed to 27 euros and fell in an almost straight line to zero until mid-2007. In the second phase, from 2008 until 2012, price fluctuations could be mitigated but were still considerable, between 8 and 33 euros per tonne of CO_2 (Committee on Climate Change, 2009, p. 68). In the carbon market that was created exactly for the purpose of climate protection, such fluctuations are a sign – especially when the price falls to zero, that is, when the market shuts down – that the cap is set inadequately. The consequences for environmental goals may be severe while those for the economy are limited to changes in production cost, of which carbon allowances represent just a fraction. In the crude oil market, however, a target that produces high price volatility has direct effects on producers' as well as consumers' behaviour and thus has a much greater potential to harm the economy. The nature of this problem is such that it cannot be resolved by simply setting the cap differently. The assessment by research or the civil society that the European emission trading system is not able to guarantee stable price patterns is basically correct (see, for instance, Carbon Trade Watch, 2013; Gilbertson & Reyes, 2009, pp. 12–13). But it is insufficient on analytical grounds. Rather than an empirical question, it is a theoretical issue. Indeed, there is no method for setting the target at a level that corresponds to stable prices. Just as lacking is the means of following a predetermined declining path of an emission cap or an oil intensity ratio, at the same time achieving a stable price (or a smoothly rising price). This is due to an endogeneity problem. For the crude oil price to be stable over time, oil supply has to accommodate changes in oil demand. Since supply is determined, or at least bounded above, exogenously by the target, it is up to policy to set it in a way that matches oil demand. On the one hand, future demand for crude oil depends on output growth. The political authority has to make predictions, which can be more or less close to the effective outcome. A prediction bias has its consequences for the oil price. If supply was market-based, producers could overcome a sudden shortage in the spot market by a decrease in inventories, which is not possible if the volume of sales is predetermined. Second, and this is the true endogeneity issue, oil demand depends on the oil price. The oil price is itself, among other factors, dependent on conditions on the supply side. In other words, oil demand is *ceteris paribus* determined by the quantity and the price at which crude oil is supplied. Supply is in turn to be determined by the political authority in light of demand expectations. We end up in a circular reference, which cannot be dissolved by the approach of the oil supply target. The decision to use an oil intensity ratio as a target instead of an absolute level of supply enables business cycles and long-run oil demand developments to at least be partially taken into account. But it cannot give a response to the endogeneity problem.

The third critical point concerns economic stability in the crude oil market. The oil supply target, if announced in public, provides oil companies with information about future production conditions. Given this transparency, they can plan production and investment in a stable environment, since they know the capacities needed to fulfil the cap. Even if the price should rise strongly in a given situation, producers do not have a reason to overinvest, since supply cannot increase despite the high price. This constraint is relaxed and relativized by a number of factors, namely foreign trade, changes in production technology and business-cycle impacts. However, stability gained in this aspect is jeopardized by other, newly emerging instabilities. First, price instability is enhanced by the theoretical inability of the political authority to exactly match oil demand, so that the oil price follows a stable pattern. Second, the issue of speculation in the futures market is not addressed. Financial investors may therefore still benefit from betting on changing oil prices. The phenomenon may even be intensified: information about the target in the future may be an additional source fuelling speculative futures market investment. For instance, given that the economy is in a boom and speculators judge the supply target to be too low to accommodate rising demand, they start betting on rising prices. By exerting additional demand power in the futures market, the crude oil price rises effectively and runs the risk of building a bubble. The effects of such a price increase have been analyzed in abundance. While overinvestment should now not be a problem of first priority, owing to the reasons explained, exacerbated oil price fluctuations have their effects on the rest of the economy.

Weighing the advantages and disadvantages of this policy proposition in a conclusion yields an ambiguous judgement. It is an effective approach to the ecological problem. However, it does not resolve the problem of economic and financial instability or even worsens it. In particular, the causal chain identified in our theoretical analysis from monetary policy to price and quantities of crude oil is not distorted in any way. Yet, the discussion of the proposal is nonetheless useful, because there are some insights that can be used in the remainder.

Fossil energy tax

A policy approach to the market for crude oil that is to guarantee economic stability and ecological sustainability at the same time needs to overcome the troubles arising from quantity regulation detected with oil supply targeting. In this section, we thus consider an approach to the ecological issue that focuses on the price instead of quantities. With respect to energy and environmental policies, an often debated proposal is a tax imposed on energy consumption, be it energy in general or crude oil in particular. First emphasized by Pigou (1920), it aims at internalizing external effects. Fully integrated in the neoclassical microeconomic framework, the analysis suggests that external effects consist of the gaps between the social marginal net product and the private marginal net product (see, for example, the summary in Aguilera Klink, 1994, p. 387). This distortion is to be corrected by a tax that, in the case of negative externalities, is imposed on the producer of this external effect. The tax is to be set such that the producer corrects its activity by an amount that re-equilibrates the social and private marginal net products. In the specific case of crude oil or fossil fuels the tax would make production more expensive, so that the price realized is higher while output is lower. In the optimum, fuel production is such that the marginal unit contributes a utility that is equal to the damage caused by additional pollution as a result of this marginal unit.

The concept drafted in this way suffers the problem of applicability to the real world. As Pigou (see, for instance, Pigou, 1951) himself states, the conception of social and private marginal products requires knowledge about utilities both at the social and individual level, which are not measurable. Yet, the need for practicability allows deviating from the exact optimal solution obtained under perfect competition. Baumol (1972, pp. 318–320) proposes a tax that is driven by one or several principal indicators concerning the externality. If the indicator exceeds a specific predetermined limit, the tax is adapted. Thereby, the political purpose is suggested to be reached step by step.

In fact, the instrument of a tax on fuel is not as outstanding as it might at first seem, nor is it merely a theoretical idea discussed in abundance but never realized. It is, in fact, quite the opposite, namely that such taxes already exist in quite diversified varieties in many OECD countries (see, for instance, Newbery, 2005, pp. 1–7). It must also be said that, concurrently there exist substantial tax preferences for producers of crude oil and other fossil fuels, for instance, in the United States (Metcalf, 2007, pp. 158–160, 166–168). Subsidies are the opposite of taxes but, in an economic sense, they are of an analogous nature. Both the subject and the object on which a tax is imposed may differ. It might be raised from the producer, the refining industry or the final consumer. Consequently, crude oil, petroleum or heating oil may be the respective object of taxation. Let us assume, for reasons that will become clear in the remainder, that the tax is raised from crude oil producers.

A tax on crude oil production, fossil fuels in general, or pollution is quite elegant in the sense that it allows for pursuing an ecological goal without being obliged to set prohibitory bureaucratic regulatory controls. The purpose can thus be reached more efficiently (Bernow et al., 1998, p. 193). An additional complication accrues when considering the economy of interest in the context of its international relationships. A tax imposed only in a single country brings losses in competitiveness for the industries concerned, since their production costs increase. Usually, these drawbacks are suggested to be overcome, on the one hand, by putting tariffs in place to protect these industries. Alternatively, the original competitive conditions can be restored by removing the tax on the goods exported and raising it on the same goods imported, to wit, crude oil in our case (ibid., p. 195).

Yet, while this correction mechanism at the country's border eliminates these distortionary effects, it does create new ones, even though they are of a transitory nature. Both measures bring about similar effects, so let us consider the case where crude oil exports are unburdened and imports are imposed by the tax. The oil industry is imposed a tax of, say, 10 US dollars per barrel. One may well imagine a regime with the tax being a certain percentage share of the oil price. However, this would imply additional dynamics, as tax revenues increase and decline with the oil price (Newbery, 2005, p. 6). We abstract from such effects for now. To speak in simple model terms, the supply curve shifts upwards, involving a shift on the demand curve in the direction of a higher price and a lower oil quantity. The oil price in the country is now higher than at the global level. Arbitrage does not lead to re-equalization of the two prices, since exports and imports are corrected by the tax. This is the standard explanation of how such a tax can be implemented in a single country against the background of global integration without distortions. However, there are cross-border effects. Under the usual case of imperfect competition, the crude oil tax is raised at the cost of producers' profits. Therefore, it becomes lucrative to export a higher share of oil production instead of supplying it at home. The supply curve thus not only shifts upwards, owing to the tax imposed, but additionally moves to the left – which is, in fact, the same direction again – owing to a larger portion of output that is exported. The oil price thereby increases further, while oil demand decreases even more. Arbitrage effects are therefore present and lead basically to a quasi-equalization of the home price and the global price of crude oil corrected by the tax (and other imperfections like transportation cost, which are ignored here). After a shorter or longer period of adjustment, we end up in a new state where the oil price in the small open economy is in a kind of equilibrium with the world price.

Adjustment bears distortion that repeats every time the level of the tax changes. Consequently, this makes it harder to gain knowledge of the effectiveness of the tax. There is a way to smooth distortions. Instead of unburdening exports and imposing imports by the whole amount of the tax, correction at the country border may only be such that it compensates for the difference between the home price and the world price. The introduction of the tax affects production costs. The new price guarantees some profits that are lower than before. It becomes more lucrative to export oil if it is fully unburdened of the tax. However, if tax is refunded only by a fraction, corresponding to the difference between the higher home price and the lower world price, the producers' profit per barrel remains the same. Exports are disburdened by $\Delta p = p_{home} - p_{world}$, while imports are charged by the same difference. There is no arbitrage opportunity between selling in the national economy and exporting and, hence, there is no incentive to raise exports in a way that would tighten the home market. Some output remains unsold at home owing to decreasing demand in response to the price increase. It can be exported without having distortionary effects in the home market, owing to the new compensation mechanism at the country's border. Once the tax changes, international trade conditions adapt smoothly, since the tax compensation adapts just as smoothly because the oil price is a part of its formula. There is no distortionary force.

Important to be mentioned, Δp may at the maximum be equal to the tax. This maximum then corresponds to the conventional mechanism of tax correction. If the difference between the home price and the world price is larger, then it is due to other reasons than taxations, such as lacking competitiveness owing to production technologies, exchange rate changes and so on. Fundamental developments of this kind cannot be taken into account by the political authority.

There is justification in asking why such a smoothing mechanism is necessary. All kinds of taxes on consumer goods and services, like value-added taxes, feature the same cross-border problem. However, this problem is usually managed by subtracting the tax completely from exports and fully adding it to imports. Nonetheless, we do not observe substantial distortions that seem to be worth an economic debate. The crude oil market is different in several crucial respects. First, it is highly globalized. Hardly any other sector is as strongly globally integrated as the market for crude oil. By contrast, many national sectors produce goods and services that are not exported and hence are not in global competition. With crude oil, the national economy and the world price are so strongly connected that changes in one place, which is the oil tax in our case, have potentially strong repercussions in the other. A smoothing mechanism is therefore important, especially as the crude oil tax appears to the market as an exogenous impact. Second, for the tax to be effective with respect to petroleum consumption, it ought to be of a considerable level. The currently existing gasoline tax in the United States of 18.4 cents per gallon (Bickley, 2012, pp. 11–12) is probably insufficient. The higher the tax is set, the larger are its potential distortions. Third, an environmental policy that seriously aims at continuously reducing the share of fossil fuels in energy consumption has to increase the tax in regular time intervals. Each change in the tax then may bring about distortions and more and more lasting uncertainty about future market developments. Fourth, in the case that the tax is imposed as a constant percentage share of the oil price, every change in the price affects the amount of the tax and thus bears additional distortionary dynamics.

While the advantages of an energy tax in general, and a crude oil tax in particular, are obvious and given by its pure motivation, the disadvantages have to be emphasized in more detail. On the one hand, there is the well-known principal critique that energy taxes would raise production costs, giving rise to a drain of production industries where fossil energy makes up a large portion of inputs. This is an argument that concerns all kinds of taxes. Yet, there are numerous other factors that make up the competitiveness and prosperity of a country (Mills, 2015). This is valid for the manufacturing industries. Crude oil producers themselves do not have to make a decision about which country to produce in. First, drilling wells cannot be moved away, implying that crude oil production is naturally tied to a certain place. Second, oil that is sold in the national economy is taxed whether it is produced at home or abroad. However, tax revenues may be used to compensate for potential losses in international competitiveness. This issue will be debated further.

On the other hand, what is interesting and more important in our context, a crude oil tax does not contribute to economic and financial stability. To simplify, it is just added to a given price level. Besides the middle- and long-run effects of the higher oil price on decreasing oil demand, the search for alternative energy sources, and technology development, the tax affects neither the fundamental nor the speculative forces that drive the oil price. The price thus is basically allowed to follow the same pattern as it does without the tax. This includes the occurrence of speculative activity potentially giving rise to price bubbles and the overinvestment resulting from it. The crude oil tax may be effective in bringing an ecological benefit. But it is not able to solve the problem of economic and financial instability. To prevent monetary policy from contributing both to instability and higher oil intensity, we need another policy approach. In the next chapter, we will develop a proposition by combining the benefits of the hitherto discussed measures. In particular, critical points will be taken into account in order to keep them out of the policy design.

NOTE

1. The public returns emerging from the auction may be redistributed to people in order
    to prevent harmful effects on equality or even to improve wealth distribution (Boyce &
    Riddle, 2007; Feasta, 2008, pp. 13–17). For the use of the revenues, see the further debate
    in this book.

6. An economically stable way out of fossil energies

To sum up, each of the previously presented policy proposals has its advantages and disadvantages. Financial market regulation by laws aims at determining, for example, speculative traders' futures positions or a tax that prevents the price deviating from its fundamental value. This presumes knowledge of the fundamental part of a price, which we suppose to be impossible. Hence, legal regulation does not necessarily bring stable market conditions, in contrast to what is intended. A constructive solution thus should be independent of the information of the fundamental and the financial market components of the oil price. Using the strategic petroleum reserve to intervene in the oil market in the presence of speculative activities is effective, but features similar drawbacks. Intervention may be overshooting, so that it decreases the oil price to a level lower than implied by fundamentals.

Setting oil supply targets to reduce the (absolute or relative) importance of petroleum in the economy is effective. But while it is able to fulfil the ecological goal, the achievement of this goal comes at the cost of additional price instability, which harms the rest of the economy by increasing uncertainty. A tax on crude oil production, by just adding another charge to the previous production costs, does not affect the oil price evolution, which is itself driven by fundamental as well as speculative developments. The impact of monetary policy on the crude oil market thus remains the same as analyzed in the preceding chapters. Hence, a tax has an ecological effect but does not contribute to economic and financial stability. Price regulation should be implemented in a way that effectively reduces price volatility.

By means of our theoretical and empirical analysis, we identified a lasting effect of speculation in the crude oil futures market on the oil spot market. Most policy answers of those authors who recognize such effects, aim at ruling out any monetary policy and financial markets effect on economic fundamentals (see, for example, Davidson, 2008; Staritz & Küblböck, 2013). To circumvent the disadvantages of the policy responses, we choose just the opposite way: instead of ruling out the impact of monetary policy and financial markets, we employ a system that is able to *make use* of this mechanism as a tool to achieve a better economic result to provide financial and economic stability as well as ecological sustainability.

What we propose in the following is an approach of macroeconomic governance implying an appropriate coordination of monetary and fiscal policy. General propositions of this type of economic governance are, for example, provided by Arestis (2015), Arestis and Sawyer (2004), Asensio (2007) and Hein and Truger (2011, pp. 214–216). The difference to our case consists in the fact that these general approaches usually focus on total output and the general price level, while we are concerned with oil quantities and the particular price of crude oil, respectively. Even though this policy is only concerned with a specific market instead of the total economy, the methodology is still macroeconomic as it relies on macroeconomic principles of monetary and fiscal policy.

6.1   THE BASIC FRAMEWORK

Achieving economic and financial stability requires a stable price of crude oil. One may argue that instead of a stable price, oil quantities should be stable in order to have stable overall economic conditions. However, considering the crude oil market as a market that (once the supply constraint is given) is determined by demand from the rest of the economy reveals that the latter requires a stable oil price. Once the price is fixed, growth in the non-oil economy determines the amount of crude oil purchased at this price. Fixing the quantity first would lead to strongly rising prices in times of high demand and thereby lead to volatile production costs and consumer prices, which again affect demand conditions. Instability would be enhanced.

Let us therefore simply argue that the oil price is determined by policy. The oil price thus becomes exogenous and the oil industry, as well as the rest of the economy, align production and consumption, respectively, with it. This allows stable conditions in the economy: supply and demand follow long-run patterns and so does investment in the oil industry. Overinvestment does not occur anymore. Since there is no oil price risk, speculative activity in the futures market becomes meaningless and hence no leverage should build up. A price bubble cannot take place anymore due to price exogeneity. This may seem quite radical. However, the debate on whether policy should let asset prices float freely or target them is quite intensive and, besides *laissez-faire* stances, numerous tools are proposed to affect asset prices (see, for instance, Bernanke & Gertler, 2001; Brittan, 2009; Palley, 2003; Williamson, 2009).

With respect to the ecological issue, the question of the right price has to be taken up again. Pure science does not provide a result of what the correct price is, since we cannot distinguish the fundamental and the financial market components of the oil price. The price is thus subject to political considerations. In order to reduce pollution, petroleum consumption has to decrease. Therefore, the price should be high. Hence, in addition to the idea of just fixing the price, we may as well say that policy decides to raise the oil price step by step in a transparent manner. Stability is guaranteed, since the price increase takes place smoothly and all actors in the market are aware of the future price development. There is thus a strong incentive for oil consumers to invest in energy efficiency and renewable energy sources. The long-term increase of the oil price guarantees that these technologies remain competitive and become even more so. Logically, a high oil price usually gives rise to increasing oil production. How overproduction can be prevented is emphasized below. Moreover, naturally, it is not inevitable for policy to raise the oil price forever. In contrast, it can define a price level that is considered as appropriate, lead the oil price smoothly to this level, and park it there.

The next two questions suggesting themselves are those about which institution determines the oil price exogenously and how it can be implemented. To address the former, we briefly go back to our theoretical and empirical analysis. We argued that monetary policy affects the crude oil market through both the real economy and financial markets, giving rise to an influence of speculation on the oil price. The central bank is in a dilemma: it may be in a situation where it is necessary to cut the interest rate target in order to stimulate investment behaviour and credit creation. However, credit is not only granted for the purpose of real production and consumption but also to invest in financial markets, including the crude oil futures market. A speculative bubble with the corresponding harmful effects on the crude oil market is likely to build up. To address both issues, that is, business cycle development and futures market evolution, it is not sufficient to have only one instrument in the form of interest rate manipulation at hand. An additional goal, to wit, oil price stability, requires an additional tool for monetary policy.

The basic idea of how the oil price target can be realized is provided by Davidson (2008): the political authority intervenes in the crude oil market by selling oil to lower its price and purchasing oil to raise it. However, as argued before, trading in the spot market with physical oil requires large efforts, since storage capacity is needed and transport has to be organized. Fine-tuning is not possible. It is therefore more appropriate to intervene in the futures market by purchasing and selling oil futures contracts (see, for example, Nissanke, 2011, pp. 54–56; Nissanke & Kuleshov, 2012, p. 35). Like this, action can be taken faster and much more flexibly. As our theoretical analysis around Figure 2.1 shows, spot and futures prices of crude oil usually are the same apart from partial effects that lead to very small and often only temporary differentials. For policy to be effective, there is hence no disadvantage if it intervenes in the futures market instead of the spot market.

The principal mechanism of how the price can be determined is as follows. The central bank offers to purchase all crude oil futures contracts at the price it has set as a target. There are arbitrage opportunities neither for producers nor for consumers nor financial investors. For example, if producers wish a higher price than the one set by monetary policy, they may try to sell their oil either in the spot market or in the futures market for a higher price. However, this is not attractive for any buyer, since a futures contract – and hence the quantity of physical oil that is represented by a contract – can be purchased at the targeted price from another producer who is indifferent about selling it to the central bank or to anybody else. Or, as another example, if an investor tries to go long at a lower price in order to benefit from a higher price differential when the contract matures, she does not find any supplier going short. The party on the short side prefers to deal with the central bank or any other trader who accepts a contract at the price target.

When the discussion concerns monetary policy and asset prices, usually it is not about targeting by direct intervention in financial markets. Rather micro- and macroprudential measures applied to bank balance sheets are proposed (see Canuto & Cavallari, 2013; Palley, 2003). Yet, targeting the oil price by futures market intervention is not a completely new approach. Since the abolition of the Bretton Woods system of fixed exchange rates, targeting the exchange rate of their own currency, has been widely adopted by central banks and is still subject to ongoing debates (see, for instance, Engel, 2010). The motives, specifically those concerning the short and medium run, are similar to those in our case: exchange rates fluctuations may lead to financial instability and encourage speculative behaviour (Filardo et al., 2011, pp. 38–40). By purchasing and selling currency reserves in the foreign exchange markets, the exchange rates can be influenced. Without judging these monetary policy objectives, they show that the approach of oil price targeting can be embedded in already existing frameworks.[1]

It is justified to ask if there is no threat of speculative attacks of financial investors who purchase futures contracts to push the oil price beyond the target. This aspect shows another weakness of Davidson's (2008) proposal of using the strategic petroleum reserve. Once the reserve is running short, the government administration runs out of its possibilities. In the case of intervention in the futures market, it is easier for the central bank to accumulate a greater stock of futures long positions, which it can use to counteract price effects of financial investment. Once these positions are reduced to zero, the monetary authority can even go short for a while to lower the oil price further. Of course, the central bank is not interested in either purchasing or selling crude oil. Hence, all positions must be offset later or rolled over to contracts of longer-lasting maturity. Thus, given a situation where the central bank holds net short positions, evening them up with long positions tends to raise the oil price again and thus benefits financial investors who just want the oil price to increase. However, the central bank is able to affect the oil price specifically and thereby can prevent investors' profits or even inflict them a loss. Such short-run threats allow the central bank to avoid attack from financial investors. It now becomes clear why it must be monetary policy to take the task of targeting and determining the oil price: the central bank is the only institution that has unlimited means to act in the crude oil market. Therefore, it is the only institution to make credible threats and hence to guarantee the realization of the target. This is a crucial difference to the use of the strategic petroleum reserve proposed by Davidson (2008).

Naturally, this approach, as it is outlined until now, distorts allocation of resources seriously, since the oil price does not react to changes in supply and demand anymore, be it in the spot or futures market. If the oil price is raised by exogenous monetary policy action, demand decreases while supply reacts, in a quite conventional way, by rising investment and growing oil production. However, the oil price does not fall by ways of the hitherto known feedback mechanism. Oil producers always have a final demander for futures contracts, that is, the central bank. On the other hand, the only way for the central bank to keep up the price at its targeted level is by purchasing an ever increasing amount of futures contracts. Since it does not want to accumulate inventories of physical oil, it has to roll over all contracts. A contract specifies the delivery of crude oil at a pre-determined date in the future, implying that it is still unsold in the present moment. Hence, stocks then build up with oil producers. They raise oil production in response to the rising oil price and thereby accumulate inventories correspondingly. Consequently, the amount of futures long positions held by the central bank has to increase proportionally. A high volume of inventories would, in normal circumstances and in the medium to long run (assuming financial market effects away), press the oil price down, since it makes the supply side of the market less tight. The only way to prevent a decrease in the price is by raising demand for these inventories. This takes place in the form of the futures contracts purchased by the central bank, which define a claim on those oil stocks.

Inventory-building may be mitigated once producers get confidence concerning the path of the oil price pursued by the monetary authority. Being aware that the central bank will never want to settle the claim on oil inventories, they might just reduce capacity utilization to save carrying costs but nevertheless contract with the central bank on the short side. Even though physical stocks are then at a lower level, oil producers are still ready to raise supply physically at the given price level if demanded physically. Hence, inventories still have the same down-pressing impact on the oil price. Monetary policy still has to intervene increasingly in the futures market. The danger of an ever-growing imbalance is thus identified: there is overproduction, which only occurs because the monetary authority subsidizes crude oil by purchasing it at the price it determines exogenously. The problem has to be met by an additional political measure.

Raising the oil price serves the purpose of reducing final oil consumption. However, owing to the relaxation of the supply side in response to oil price growth, we end up with overproduction. Therefore, we are back to the initial reflections in this chapter, where the question of the optimal price was asked. A continuous increase in the oil price is ecologically sustainable from the demand point of view but brings unintended consequences on the supply side. In order to decrease the oil intensity in the long run, a policy is needed that decreases oil consumption as well as oil production. There is a way to achieve this: if oil production is charged with a tax, production costs increase and the oil supply curve shifts upwards or to the left. At a given oil price, oil supply is lower than it would be without the tax. Except in the very short run where the supply curve is vertical, the tax is effective in influencing the supply side in the pursued direction whether the supply curve is upward-sloping or horizontal in the medium to long run.

Yet, some drawbacks of an energy tax have been outlined above. The principal one is that a tax does not rule out speculative influences on the oil price and hence does not contribute to price stability. In our proposal, however, we can eliminate these shortcomings if we combine the tax with the price target. The cooperation of monetary and fiscal policy becomes the central issue. The tax has to be set in a way that it matches the difference between pre-tax oil production and oil consumption. To put it analogously, the tax should be imposed so that it avoids the accumulation of inventories. Such a system prevents the central bank from being obliged to purchase crude oil futures in an ever increasing volume. The challenge consists of the optimal calculation of the tax rate. Clearly, it has to be flexible over time to account for shocks occurring in the crude oil market. The level of the tax should be aligned with the evolution of oil inventories. When oil inventories increase, oil production is too high with respect to demand at the given price, implying that the existing tax level is too low. Conversely, when oil stocks decline, the tax is too high, because demand is larger than production at the given price. In the following period, the tax should be adapted in the corresponding direction.

In practice, however, oil inventories are inappropriate to be taken as the target variable that determines how the level of the tax should be set. As discussed in several places above, inventories are quite difficult to measure owing to unsatisfying data availability. Raising data would take time, so that they would only be available with considerable time delay. Moreover, inventories do not necessarily need to evolve as suggested by neoclassical theory. There is no reason why production capacities should be used fully or at least at a constant rate. Stocks then do not accumulate at a speed at which they would in the case of constant capacity utilization, but the problem of overcapacities that put downward pressure on the oil price remains.

It is therefore much more adequate to use crude oil futures held by the central bank as the variable that is to be targeted. They are the relevant issue, since they reflect the state of the crude oil market better than oil inventories: their level incorporates the states of, both registered and unnoticed, inventory accumulation as well as capacity utilization. Moreover, they are of broader public interest, since it is their amount that determines to which extent oil price targeting leads to a subsidy of private oil producers. When the tax is too high, demand is higher than production at the given price level, so that the central bank has to go net short in order to prevent the oil price from rising higher than the target. In an analogous way, when the tax is too low, production is higher than demand and hence monetary policy has to purchase futures long positions in order for the oil price to be kept at the target level so that it does not fall below it. This action guide for the central bank may appear as just the same as if oil inventories were targeted directly. But besides the drawbacks of oil stock data owing to data quality and capacity utilization, targeting oil futures instead of oil inventories has an additional crucial advantage for practical use: futures data feature much less time lags, since trade in the futures market is even reported in high-frequency data (although the latter is, yet, not necessary for the implementation of the price targeting system we advocate here).

There is another advantage that makes this policy proposal practicable. It is reasonable that the oil supply curve is either rising or horizontal in the middle to the long run, while the demand curve is falling over the same time span. However, neither do we know the exact slopes nor are they constant over time. Specifically in the short run, the curves can basically have any slope, leading to radically indeterminate outcomes (see Pilkington, 2013; Varoufakis et al., 2011, pp. 294–298). Investigating the crude oil market in this respect would require hard econometric analysis with probably imprecise results. However, radical indeterminacy or radical uncertainty of the crude oil market is not a threat to oil price targeting, because it is already incorporated in the amount of futures contracts that the central bank has to purchase. Assume that a specific event leads to an increase in the precautionary demand of consumers who expect the crude oil price to climb in the future. Inventories decrease (or capacity utilization increases) and the amount of futures contracts that oil producers deal with the central bank declines. Oil producers may react by keeping crude oil from the market to raise the price further. The monetary authority then has to counteract by raising its short positions so that the oil price does not exceed the target. Net long positions therefore fall. Analogous shifts occur with changes in fundamentals, be it a technology shock raising supply or accelerating economic growth that leads to higher demand. In all cases, the changes are reflected in the account of central bank futures holdings. Every time such an event takes place, monetary policy has to defend the price target by trading futures contracts in the first step. In the second step, it has to change the tax on oil production so that the account of futures can be kept constant. Of course, futures holdings are allowed to fluctuate over time, since, owing to uncertainty, it is not possible to assess the tax perfectly so that the central bank does not have to intervene in the futures market anymore. Yet, this is not a grave problem, since the central bank has unlimited purchasing power and can afford fluctuation in futures positions. It is in the middle to the long run that futures holdings should be constant in order for the crude oil spot market not to become structurally imbalanced. Such intervention is possible without the knowledge of elasticities and short-run and long-run dynamics in the crude oil market.

This system can briefly be described by saying that the price target is the exogenous variable while the tax is set exogenously as well but has an endogenous meaning, because its optimal size is determined by other market forces. One may argue that it is the tax that has the crucial impact on the price rather than trading with futures contracts by the central bank, because contracts do not increase in the long run and therefore cannot have an effect on the price. This is basically true. However, our approach requires the procedure described here. It is price targeting and futures trading that guarantee a smooth price and thereby economic and financial stability at first instance. At second instance, it is the tax on oil production that rebalances the oil market. If only a tax were imposed, the oil price would fluctuate as it would without any political intervention and there would be no hint of the level at which the tax rate should be set.

This two-stage implementation of the oil price targeting system outlines the political organization necessary for it. First, it is monetary policy that sets and realizes the oil price target. It observes the level of oil futures contracts in its account and hence provides an advice of the level at which the oil production tax should be set. Second, fiscal policy implements the tax at the proposed level.

An additional issue concerns the question whether the oil price targeting system affects the other goals of monetary policy negatively. A steadily rising oil price passes through to inflation even though the estimated effect is found to be limited and to have decreased in past decades (see, for instance, Cavallo, 2008; Cecchetti & Moessner, 2008; Chen, 2009). Nonetheless, one might fear increasing inflation owing to this policy proposition. However, given that price changes from year to year are moderate, the impact on inflation rates is even more so. All in all, oil price targeting may even yield a benefit for monetary policy. Firstly, even though the oil price increases, it does so in a smooth way. In contrast to the volatile oil price pattern of today, which has a corresponding effect on fluctuations in inflation rates, oil price targeting may slightly increase the average rate of inflation but reduces inflation rates volatility. Moreover, of course, the oil price does not necessarily have to rise forever. Once an appropriate level is reached, the price can be parked there. A constant and stable oil price then definitively does not accelerate inflation. To sum up, oil price targeting affects inflation targeting to the extent that the former creates an 'oil price transmission channel', which transmits monetary policy action to the general price level. Its importance is an empirical question that suggests it to be quite limited. On the other hand, the oil price transmission channel may from time to time even be a useful tool for monetary policy to stabilize inflation rates.

Further concerns may arise pointing at the fear of inflation in the course of rising money supply owing to futures purchases. This is the monetarist idea that growing money feeds more or less directly into higher prices (see Mishkin, 2006, pp. 2–4). Yet, money is still endogenous. The central bank does not raise money supply by its own force, which then leads to a rise in the general price level. Rather, it sets the oil price target. At this level, the quantity of money is a result of the number of futures traded with the monetary policy. To be exact, the largest part of the purchasing power created by futures trades does not take the form of official money but rather of a large leverage, as explained in our theoretical analysis. Moreover, since the futures account of the central bank is targeted to be stable, there is no rising quantity of money by an ever increasing number of contracts to be expected. Money demand increases to the extent that more money is needed to make transactions with crude oil that is now more expensive. Still, money is the result rather than the cause (in this regard, see Davidson & Weintraub, 1973). Hence, any influence of the oil price targeting system on inflation is given by the changed and smoothed price pattern of crude oil. Monetary concerns are misconceived.

For the balance sheet of the central bank, there is no risk of loss that the public is charged with. If the tax on oil production is set adequately, the account of crude oil futures at the central bank does not increase in the medium and long term, so that no systematic asset price risk emerges. If anything, there is a benefit for the central bank, because once the number of futures can be kept more or less constant, the rising price of each futures long position yields a return. For a volume of positions sufficiently low, they could even be liquidated without having a lasting impact on the oil price.

It should not be denied that a rising oil price is a challenge for many industries as they need to change production technologies and sometimes even have to develop new ones. However, tax revenue may be redistributed so that the tax is neutral with respect to production cost from a macroeconomic point of view. Importantly, if the tax fund is redistributed to oil companies, it should be earmarked in the sense that it must be used for the development and production of sustainable technologies or the exploration of renewable energy sources to avoid any moral hazard. Otherwise, producers would have an incentive to increase crude oil production as much as possible, since their income from tax distribution would grow accordingly. The tax would lose its impact.

6.2   OIL PRICE TARGETING IN THE SFC MODEL

After having outlined the basic working of the oil price targeting system, the use of our SFC model helps to look at it in some more detail. The model is taken in the same form. Step by step, it is modified and slightly extended to introduce the oil price targeting system.[2] First, the futures price target of the central bank is defined as an exogenous variable. Since the central bank has unlimited capacity to reach its target by trading crude oil futures, the target can be set equal with the actual futures price.

$$p_{fut} = p_{fut,\,target} \qquad\qquad (6.1)$$

Naturally, this is a simplification since, in practice, the monetary authority may fall short of reaching the target in the very short run owing to erratic fluctuations. However, by appropriate reaction, the central bank can approach the target.

Oil producers' profits are now diminished by the production tax imposed on them.[3] Production profit equation 4.8 is thus modified in the following way:

$$PP_P = Y - C_d - W_d - r_{-1}*L_{P,-1} - T \qquad (4.8')$$

where T is the total sum of the tax paid by producers. Since the tax lowers profits, future profit expectations are also downgraded, which reduces expenditures for real investment. The latter can also be negative, thereby representing the shutdown of production facilities. The resulting shrink in the capital stock implies reduced production capacities. The oil price rises in response. An additional and more immediate price-rising effect of the tax takes place through increasing production cost. We model this fact by adapting model equation 4.16 to become as follows:

$$p_{fut} = \frac{\delta_4 + \delta_5*(F_L^{tot} + C_{oil,d}) + \delta_8*T}{\delta_6*K_{-1}} \qquad (4.16')$$

Equation 4.16' shows that the higher the tax, the higher is the price. δ_8 determines the extent to which the tax affects production costs. There could as well be the spot price on the left-hand side of the equation instead of the futures price. However, since equation 4.16, and hence equation 4.16' as well, represents the integration of the spot and futures market, and since we had set both prices equal in equation 4.15, this does not matter. As another modification, F_L has become F_L^{tot}. Before, private financial investors were assumed to be the only actors to go long in the futures market. Under the oil price targeting regime, the central bank becomes a futures dealer, too. Yet, what is relevant is not only speculators' positions but also all long positions in the market, that is, F_L^{tot}, because all futures have the same effect on the price.

Now, the futures price is defined endogenously twice, that is, in equations 6.1 and 4.16'. Since the futures price is given by the realization of the exogenous price target, it enters equation 4.16' as a predetermined variable. The variable that is determined endogenously in this latter equation is total futures positions. Reformulation yields:

$$F_L^{tot} = \frac{p_{fut}*\delta_6*K_{-1} - \delta_4 - \delta_8*T}{\delta_5} - C_{oil,d} \qquad (4.16'')$$

The intuition is now in line with the above argument: the higher the price (which corresponds to the price target), the larger the volume of total futures positions has to be *ceteris paribus*, so that a lower spot oil demand,

$C_{oil,d}$, does not pull the oil price down. However, there are different counteracting effects. First, a higher capital stock requires even larger futures positions, since overcapacities put downward pressure on the oil price. Second, the tax on crude oil production works in the opposite direction. The higher the tax, the stronger is the supply constraint in the spot market and the stronger is upward pressure on the price. Consequently, the less demand for crude oil futures is required for the oil price to stay at the targeted level.

The fact that there are now two different types of futures contract traders, to wit, private financial investors and the central bank, presupposes two other simple adjustments in the model. On the one hand, 4.17 concerns now not total futures long positions but only those of private financial investors, F_L^I. This is only an adaptation of denomination. On the other hand, total futures long positions are composed, naturally, by private financial investors' and central bank exposures. Central bank positions are therefore calculated as the difference between total and private positions:

$$F_L^I = \frac{M_I}{m * p_{fut,-1}} \tag{4.17'}$$

$$F_L^{CB} = F_L^{tot} - F_L^I \tag{6.2}$$

The maintenance margin that the central bank has to pay, assuming that it participates as a conventional trader without any privileges in the futures market, is then simply given by the multiplication of the rate of margin requirement with the price level (to which the requirement rate is related) and with the volume of futures positions. It is analogous to equation 4.17', but reformulated. Furthermore, total maintenance margins deposited with the banking system to which, to remind, we assume the futures market clearing house to belong, is the sum of financial investors' and the central bank's margin:

$$M_{CB} = m * p_{fut,-1} * F_L^{CB} \tag{6.3}$$

$$M_B = M_I + M_{CB} \tag{4.31'}$$

Equations 4.16'' and 6.2 show that the central bank is ready to step in and buy the amount of futures necessary so that the total of contracts in the market is such that the futures price matches the targeted level. The tax now should be set so that the central bank does not have to go excessively net long or net short to realize the oil price target. Moreover, for practicability, the monetary authority should be able to derive the level of the tax by means of reliable and observable indicators. In the model, we define the tax to depend on two indicators, namely the oil price target and the central bank futures positions::

$$T = i*(\tau_0 + \tau_1*(p_{fut,target} - 2) + \tau_2*(p_{fut,target} - p_{fut,target,-1}) \quad (6.4)$$
$$+ \tau_3*(F_L^{CB} - F_L^{CB,target}))$$

where i is a policy dummy that takes the value 0 before the oil targeting system is introduced and the value 1 thereafter. τ_1 measures the importance of the level of the price target for the tax to be set. The number 2 signifies the crude oil price before policy intervention. Hence, the numerical value does not have any further meaning but is just given by model calibration. τ_2 reflects the impact of a change in the price level. And finally, τ_3 shows how much the tax should increase (decrease) if the central bank futures account increases (decreases) as the result of the defence of the price target. Here, the central bank may define another target concerning the amount of futures holdings. It might be set to zero, implying that the central bank adjusts the tax in a way that it approaches zero futures holdings. However, it might be better to target a position level above zero, that is, net long. Like this, the central bank has a larger range within which it can act. It then has the possibility to have a lowering impact on the oil price in a given moment without having to go net short. While the first two terms of equation 6.4 represent the basic pattern that the tax has to follow in the face of a rising price, the third term is the short-run guide to monetary policy. It incorporates fundamental and other shocks that affect the oil market and thus crystallize in the central bank futures account.

Obviously, equation 6.4 resembles a kind of Taylor rule (see Taylor, 1993). Indeed, in order to model the behaviour of the central bank, such a mechanic formula is inevitable. This is necessary, on the one hand, because every model requires simplification to be a model. And it is possible, on the other hand, since once model parameters are given, every shock of the same type and the same magnitude has the same effect. In reality, however, the impacts of shocks differ depending on specific historical circumstances. A fixed formula may therefore be helpful in one moment but inappropriate in another. The critique of the Taylor rule (see, for instance, Rochon, 2004) applies to equation 6.4, too. It should therefore just be seen as a model equation rather than as a reliable monetary policy rule. Reality is too complex in order for the central bank to always respond in the same way to changes in its futures account. We will see that even in the model, different shocks have different optimal reaction equations.

The running of the model again takes place in the arbitrarily chosen time span from 1990 until 2014 at weekly frequency. Assume now that the central bank decides to target the crude oil price from the beginning of 1994 onwards. It raises the crude oil price by 0.1 per cent in each period until end 2005. Thereafter, the price increase flattens and slowly converges to a new level. Meanwhile, the tax on oil production is needed to keep the oil market in balance. The targeted level of the futures account, $F_L^{CB,target}$, is assumed to be zero for now. Figure 6.1 shows the consequences for a set of variables. Again, the numerical values originate from model calibration and are arbitrary. The pattern of the oil price in panel (a) shows that the total price increase amounts to about 120 per cent. This may seem extreme. But a doubling of the oil price within 17 years is not at all unrealistic. In contrast, history has featured much stronger price changes. Panel (b) exhibits the path of the tax on oil production, which goes straight and upwards. It is just in line with what is expected from the preceding draft of the system. In panel (c), the central bank futures account shows that the tax is effective: it allows the central bank to target the oil price and to keep the account of futures positions quite stable at the same time. Note that futures are not exactly equal to zero. This shows that even in a theoretical model short-run fluctuations have to be accepted. However, they are quite small and move closely around zero. Panel (c) can be set in relation to panel (a) of Figure 4.1, where the effect of a change in the interest rate on futures market speculation is shown. The amount of futures held by financial investors is a multiple of what the central bank has to trade to meet the oil price target. An analogous result is exhibited by crude oil inventories in panel (d). They fluctuate around zero (which we have to read as a change in relation to initial stock holdings rather than positive and negative stocks in absolute numbers). Again, they are quite small compared to the amount of inventories accumulated in the face of speculation in the crude oil futures market (see Figures 4.8 and 4.9).

The further-reaching effects of the oil price targeting system that occur in the spot market are presented in panels (e) and (f) of Figure 6.1. Deteriorated profit perspectives lead to disinvestment, that is, shut-down of drilling wells and other production facilities. Once the oil price target grows at a diminishing rate, disinvestment slows down as well and approaches zero again. As a consequence, the capital stock in panel (e) that represents the state of production capacities drops first and then converges to a new and lower level. To see how the ecological purpose is affected, we look at the final indicator, that is, crude oil consumption in panel (f). It decreases smoothly and then converges to a lower level. Thereby, a smooth pattern is given for the economy to move out of fossil fuel dependence without creating imbalances. Analogous to the supply side of the oil spot market, the demand side faces a stable long-run price pattern. There is no

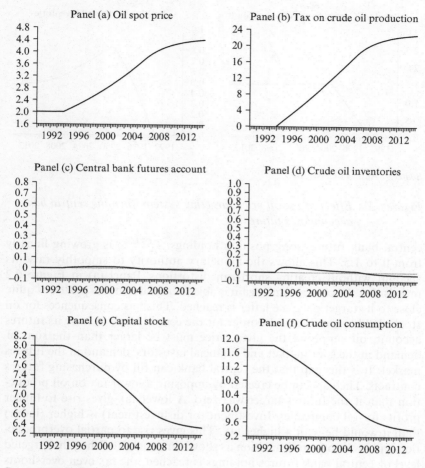

Source: Author's elaboration.

Figure 6.1   Effect of an increasing oil price target

reason for precautionary oil purchases by consumers since there is no price uncertainty. They may want to make large purchases when the oil price is still low. But the profitability of such strategies is quite low as storage is costly, storage capacities are limited and the price increase, which is a condition for precautionary purchases to be profitable, is very slow.

Alternatively, the central bank may not aim at holding no futures at all but might rather set a target level above zero in order to have a broader range for short-run intervention. For example, assume that the target of

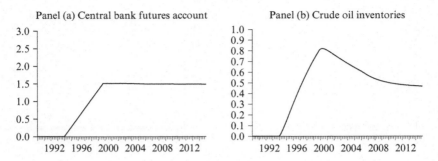

Source: Author's elaboration.

*Figure 6.2   Effect of the oil price targeting system when the central bank
            accumulates futures*

central bank futures long position holdings, $F_L^{CB,target}$, is growing linearly
from 0 to 1.5. This allows the monetary authority to smoothly raise its
account without creating short-run distortions. Panel (a) in Figure 6.2
reveals that the central bank again is able to hold its futures positions quite
close to its target once the latter is reached. This has consequences for oil
stocks held by producers. In order for the central bank to raise its futures
account, oil supply at the given price must be larger than the sum of
demand in the spot market and financial investors' demand in the futures
market. It is this gap that the central bank can fill by purchasing futures
contracts. The gap can be created by imposing a lower tax on oil production than if the futures target was zero. A lower tax gives rise to higher
profits for oil companies. Investment (or disinvestment) is higher (lower)
than it would be with a higher tax. This gives rise to partial overproduction and inventory accumulation as shown in panel (b). When the targeted
level of central bank futures holdings is reached, the tax even overshoots
a little, owing to time lags in the reaction function. Oil stocks fall back but
converge to a positive level that corresponds to the futures account of the
central bank. This means that the central bank possesses the claim on these
inventories in the form of the futures contracts. If there were developments
in either the futures or the spot market that tend to push the crude oil price
above the target, the central bank could counteract this threat by selling
long positions or by acquiring short positions. This would raise oil supply
and thereby lower the oil price. If the contracts purchased by private
traders are not rolled over but settled, inventories would *ceteris paribus*
decrease correspondingly.

   Until now, the oil price targeting system is tested against the background
of a stationary economy without growth, changes in technology or even

shortages in supply due to exhausted sources of crude oil. It is, of course, an important issue to investigate how the oil price targeting system can be sustained when there are shocks to economic fundamentals. As an example of such a case, assume that after the oil price target has been set and implemented from 1994 onwards, the economy starts growing by an annual growth rate of 2 per cent. After 2002, GDP growth decelerates and the economy converges again to stationarity. This is an appropriate way to model a demand shock. The growing variable is again non-oil output, C_s, which approximates total output fairly well. The target of central bank futures holdings is assumed to be zero.

The variables are shown in Figure 6.3. The oil price in panel (a) still follows the same path since the target of the central bank is unchanged. The tax on oil production in panel (b) is slightly lower when the growth period sets in. The relative decline in the tax rate relaxes conditions on the supply side that are necessary to meet additional demand at the given price level. The appropriate tax charge allows for a rather stable central bank futures account in panel (c). In contrast to the stationary case, however, it grows a little before it can be kept stable. Even though the outcome is considerably positive, time-specific action of the central bank may improve it. This shows that, in reality, the central bank should not act in the mechanical manner of a reaction function that we are forced to employ in this model.

The pattern of inventories in panel (d) is now not completely analogous to central bank futures holdings anymore. Before economic growth starts, oil stocks move close to zero and are identical to the stationary case. Thereafter, however, they fall clearly below their initial level. This is not due to a mismatching monetary policy but is rather a result of economic growth: when the oil price starts increasing, production capacities are still the same as in the previous periods. Hence, to satisfy oil demand, inventories have to be released. Even though capacities are increased in response, an observation yet to be explained, the decline in stocks continues. Capacity extension is lagged, so that growing oil demand has to be met by an additional depletion of inventories in every period. Once economic growth converges to zero, inventories tend to stabilize aside from the small and flat variations that were already present in the stationary case.

The capital stock evolution is mirrored in panel (e). It declines owing to falling oil demand caused by oil price targeting. Economic growth causes a temporary re-increase: it raises oil demand and thus producers' profits. They enlarge production capacities. Once growth fades out, investment turns again into negative such that production capacities decline further to the new long-run level. The final capital stock is, however, larger than with stationarity since GDP growth leads to a higher final economic output.

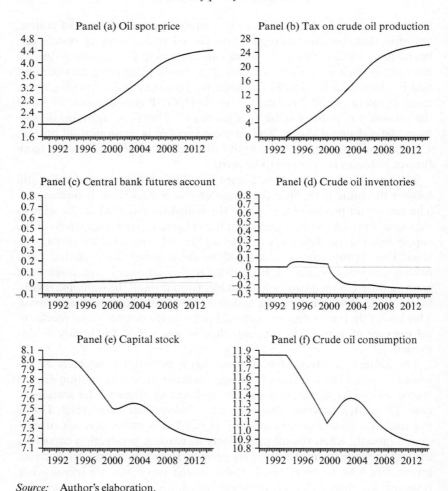

Source: Author's elaboration.

Figure 6.3   Effect of an increasing oil price target in the presence of economic growth

Crude oil consumption in panel (f) features a rather similar development. Yet, it is less smooth since demand reacts immediately to a change in the oil price or in non-oil output in contrast to the capital stock, which changes with time delay. It shows that the oil price targeting system does not rule out market mechanisms. Nevertheless, it is able to reach its ecological purpose. Absolute crude oil consumption rises in the course of economic growth. However, over the period of, say, a whole business cycle, it declines. Relative oil consumption, that is, the oil intensity of the economy, decreases continuously over the observed time span.

It is important to mention the issue of financial and economic stability. Oil price targeting ensures stable conditions for the rest of the economy in what concerns oil supply. For the oil industry, one may consider fluctuations in inventories, investment and oil consumption as a sign of instability. First, however, these fluctuations are not caused by policy intervention. They enter the stage whether the oil price is targeted or not. Second, monetary policy can hold the oil price stable according to its target despite shocks to fundamentals. Hence, while both oil consumption and the oil price would vary without intervention, intervention can at least hold the price stable. Thereby, only consumption is left fluctuating. It is true that consumption volatility is greater than otherwise since there is no feedback mechanism from a rising oil price. However, for the non-oil economy, stability is enhanced.

Concerning financial and economic stability within the crude oil market, the oil price targeting system does not allow speculation to have a price effect, so that harmful distortions in crude oil production and consumption are ruled out. There is therefore no reason for financial investors to expose themselves to risks. Economic stability is established as well. Naturally, there are fluctuations in investment behaviour of oil producers. Yet, such investment is just natural, because it is a response to shocks in economic fundamentals and thus to increased needs by the rest of the economy. Oil produced by thereby created additional capacities serves the physical needs of oil. It is the result of resource allocation in a market economy, be it perfect or not. In this sense, such investment is not overinvestment.

6.3   GLOBAL IMPLEMENTATION

The hitherto discussion and the SFC model considered a closed economy. Translated to the real world, this implies a global implementation of the oil price targeting system. Indeed, this would be the most effective solution. It requires an international agreement on the design of the system. Oil price targeting is most easily conducted by the US Federal Reserve, because crude oil is mostly traded in US dollars and the best established crude oil futures contract is the WTI Light Sweet Crude Oil (Fattouh, 2011, pp. 54–55). The tax on oil production is directly imposed on oil companies in different countries. If national governments were themselves responsible for tax collection, there would be a moral hazard problem for all countries where oil producers are owned by the state. A higher oil output yields higher tax income. There would thus be an incentive to raise oil production, which would make the tax useless. Hence, there should be an international institution able to impose a tax. Redistribution should be such that oil-producing countries are compensated – especially developing countries – for escaped benefits. However, as already noted above, it must not create a new moral hazard.[4]

6.4   IMPLEMENTATION WITHIN NATIONAL BORDERS

Global implementation requires many preconditions, such as the existence of a strong multinational institution. Since important conditions are not fulfilled, the oil price targeting system must be applicable to a single country in order to be viable. Now, it is shown how the system may be drafted for either oil-exporting or oil-importing countries. Let us start with the case of an oil-exporting country. Its central bank determines the oil price by intervention in the futures market. Oil producers at home are taxed at the required rate. Leaving the system like this would raise a twofold problem. First, consumers purchase crude oil from abroad, because the price at home is higher than the international price. The intended reduction in crude oil consumption then cannot be achieved. Yet, at the same time, demand for oil produced in the national economy is immediately zero owing to arbitrage behaviour (if we ignore transportation costs and time lags). This gives rise to economic instability within the crude oil market. Second, the oil companies would reduce supply, too, because the tax leads to lower profits and thus disinvestment even if the demand shift did not occur. This is just what is wanted by policy. However, a strong reduction in supply of an oil-exporting country has respective consequences for the world market to the extent of the country's importance. The world price of crude oil shoots up, leading to economic instability in the rest of the world. On the one hand, the non-oil economy is concerned. On the other hand, other oil-exporting countries would raise oil supply to counteract the oil price increase, so that a strong oil price volatility may be the final result.

   For these reasons, a mechanism at the country's border is required to allow for a smooth reduction in oil consumption in the national economy without distorting the world market. The easiest way to do this is the one described above with the oil production tax: when oil is exported, the difference between the home price and the world price should be transferred to oil producers. If so, then the national economy is decoupled from the world market and the oil price targeting system can exert its effects within country borders. Oil consumers are indifferent between purchasing oil at home or abroad, while oil producers may export the oil that is not consumed at home under the same conditions as before the introduction of oil price targeting. The remuneration difference again must not be greater than the tax itself. If differences between the world and the home price of oil still exist after maximum remuneration, they are due to fundamental differences in oil production costs and, hence, competitiveness. It is not the task of policy intervention to account for them in this case.

Even though a fraction of tax income has to be remunerated owing to exports, net tax income is always equal or larger to zero:

$$T*C_{oil,s} - min(p_{home} - p_{world}, T)*(export - import) \geq 0 \qquad (6.5)$$

The tax per barrel of oil produced times oil output yields total gross tax income. The difference between the home price and the world price of oil is usually smaller than the tax per barrel. Otherwise, it is the tax per barrel that can be remunerated at the maximum. For an oil-exporting country, oil exports are always larger than oil imports. But net oil exports are usually lower than oil production, since some part of oil is consumed at home. Thus, in practice, tax income is rather strictly than weakly larger than zero. This is important, since additional tax income can be used to support industries that threaten to change their offices and production facilities to other countries where energy costs are lower. Moreover, fairer wealth distribution in society can be achieved by (partial) tax redistribution to consumers (Boyce & Riddle, 2007) and renewable energies as well as energy efficiency may be supported.

In oil-importing countries, the oil price targeting system can be introduced, too. Once the central bank of the country has set the oil price target above the world price level, every importer faces an arbitrage opportunity by selling crude oil to the central bank instead of a final consumer. This will last until the crude oil price meets the price target. The tax has to be imposed on crude oil imports to prevent the situation where more and more oil is imported and that the central bank needs to purchase an ever-growing amount of crude oil futures in order to keep the oil price at its target. Now, the level of the tax is exclusively determined by the difference between the home price and the world price of oil. Tax income is strictly larger than zero:

$$(p_{home} - p_{world})*(import - export) > 0 \qquad (6.6)$$

owing to both factors being greater than zero. Again, tax income might be used to support consumers and the industries most affected by higher energy costs.

Until now, the case seems to be clear that the country described is the

United States, because crude oil trade is denominated in US dollars and the most important futures exchange is the NYMEX. If every country in the world or every currency area with a central bank should be able to adopt oil price targeting, the system must not depend on the US dollar. A short outlining shows that this policy proposal basically can work with whatever the currency is in the country considered. If there exists no futures market in that country, yet, the central bank can nevertheless contract on future oil deliveries, that is, it can 'issue' futures. Even if the market's liquidity is low, the arbitrage opportunity created by offering a higher price than the existing price level will do its job. The price target can therefore be realized. There are now two possibilities to achieve the purpose. The central bank may trade these futures contracts in its own currency. Oil price targeting and the setting of the tax on crude oil production then has to take into account changes in the exchange rate of the home currency *vis-à-vis* the US dollar in order to prevent arbitrage exports and imports. Alternatively, the monetary authority may trade the futures in US dollars by using currency reserves.

The concern that foreign oil producers and financial investors may make profitable use of the central bank's intervention activity is unjustified. As for speculators from abroad who are interested in benefitting from price increases generated by the home central bank, the same applies to financial investors originating in the national economy: the higher level of the oil price compared to the world price does not make any difference in this respect. What is relevant for speculative profits is the price change over time, which is given by the central bank. With oil companies that want to contract on the short side with the central bank to be able to sell at a higher price, there is no way to make extra profits, either. Once the contract is fixed at the oil price that the central bank has determined and should be settled, the seller pays the tax when he exports oil to that country. Since the tax is, in this case of interest, exactly equal to the difference between the targeted price and the world price of oil, there is no higher profit for the oil producer than if he would sell it in the rest of the world at the world price.

6.5   CRITICAL ARGUMENTS

It is not possible to draw a policy design here in great detail. The oil price targeting system is a political idea that requires more elaboration. At this stage, some counterarguments and assumed failures of this approach should nevertheless be addressed. Let us show that they are either misconceived or, if they are justified, how they can be taken into account by modifying the approach.

- *Feasibility:* there is a relationship of power between the central bank and private futures market traders. This is not exceptional but rather usual when monetary policy has to defend a target by direct market intervention. The more credible the central bank can communicate to defend its target, the less likelihood of potential attacks. We already mentioned the case when financial investors try to push the price beyond the target by purchasing futures long positions. The central bank may use its futures reserves and let the price fluctuate. This accrues losses to financial investors who cannot sustain them over a longer period owing to the high leverage contained in futures deals. The central bank should be able to hold off such speculative attacks. However, there is another possibility how the power relationship between the central bank and other agents may become relevant. Imagine that the contracts held by the central bank roll out in a specific moment. Since it has no interest in possessing crude oil physically, the central bank aims at rolling over its long positions. Assume further that neither are the dealers on the short side willing to continue contracting nor is anybody else in the market ready to take over the short positions. The motivation may be that oil producers want to lower their inventories to save storing costs and to realize a profit instead of only having the value of crude oil in its material and illiquid form. Is the central bank now obliged to purchase physical oil and to store it somewhere? It is not, if its power of intervention against such behaviour is sufficient. The central bank may just resell the futures positions to another agent in the market. If, as we assume, there is not any consumer or financial investor to buy it at the given price, the monetary authority may simply offer these futures at a slightly lower price. There is now an arbitrage opportunity for traders to purchase the long positions of the futures offered by the central bank instead of those at the otherwise higher market price. The central bank's action lowers the crude oil price market-wide and thereby devaluates all existing inventories held by oil producers. Moreover, future profit prospects are deteriorated should the price development take place at a lower path than it would otherwise. There is, on these grounds, no reason for oil producers to challenge the central bank by not rolling over the futures contracts. On the other hand, to mention it again, it is the task of monetary policy to apply its instruments – price targeting and the tax on crude oil production – appropriately to avoid excessive futures accumulation with the central bank's account.
- *Dealing with speculative waves:* there may be the risk that the central bank runs out of its futures reserves when it has to counteract a

speculative wave that pushes the oil price upwards. There are several
responses to such a situation. To keep this risk low *a priori*, the oil
price target changes from year to year should it not produce more
attractive yields than those of the risk-free interest rate, say, the
returns of government bonds. Like this, there is no specific reason to
invest in crude oil futures, since the path of the price is determined
exogenously and does not yield more than government bonds. This
does not mean that the percentage price change has to be exactly
equal to the risk-free rate at the maximum. There are transaction
costs and market imperfections making arbitrage imperfect as well.
In addition, risk-free rates may vary over time, thereby contributing
to uncertainty. Furthermore, the development of the crude oil price
is not completely risk free. As described just before, the central bank
may choose to let the oil price deviate from its target in the short run
to hold off financial market price effects. In this respect, it may be
useful not to target a single price level but rather a price band within
which the oil price is allowed to fluctuate. Going even further, the
central bank may set a target but allow for uncommunicated small
fluctuations around it. This strategy is related to the concept of
'constructive ambiguity' that is debated in many fields of monetary
policy (see Chiu, 2003). In the case that a financial market effect lasts
a longer time and outlives those counteracting measures, the tax on
oil production and the remunerations on exports and imports of oil
may be adjusted. In times of an oil price that threatens to be driven
upwards by financial investment, the central bank may lower the tax
that relaxes supply conditions. Open interest in the futures market
increases and since it is driven by supply it helps keep the oil price
at its target. All these instruments can be applied quite flexibly and
hence should be able to prevent large distortions owing to futures
market investors.

- *Strong fundamentals shock:* as a similar scenario, there may be a
demand shock in the spot market that is due to strong global economic growth. The central bank may react by lowering the tax and
its futures account. However, the shock may be so large that it still
continues when all oil reserves are exhausted. This case is unlikely,
because the central bank has time to accumulate oil reserves on
the one hand. On the other hand, a fundamental shock is now not
strengthened by speculation as it would probably be otherwise. But
if this scenario realizes, nevertheless, the central bank may suspend
oil price targeting for a while and only intervene if further price
increases are found to be more due to futures market investment
than to the real economy. The central bank may re-establish the

target in due time. Such partial free floating gives rise to instability. Yet, it is only of a temporary character and just reminds of the volatility that exists without any policy intervention.

- *Substitution of other fossil fuels:* when the oil price increases continuously owing to targeting by monetary policy, oil consumers may easily substitute other energy sources for petroleum. The most important substitutes are coal and natural gas, to wit, fossil fuels as well. The goal of pollution reduction is missed. Substitution of natural gas is an ecological improvement while substitution of coal implies an aggravation: the burning of a BTU of petroleum produces more carbon dioxide than that of natural gas but less than that of coal (EIA, 2015a). The cointegrating relationship suggested above does not hold anymore, because the oil price is determined exogenously. There are still endogenous responses from the coal and the natural gas markets to changes in the crude oil price. But the reverse does not hold anymore. The solution to this problem lies in an analogous proceeding with coal and gas. Their financial asset forms as futures contracts can be held by the central bank in order to target coal and gas prices. They exist already at present, for example, at the NYMEX (NYMEX, 2015).

- *Central bank independence:* there is a great body of literature about central bank independence. The proponents of a high degree of monetary policy independence from governments argue that independence allows central bankers to pursue the goal of price stability without being constrained by other competing purposes (Alesina & Summers, 1993, pp. 151–152). Empirical evidence as to whether independence effectively reduces inflation rates is both found to be positive (see, for instance, Alesina & Summers, 1993; Cukierman et al., 1992) and negative (see, for example, Campillo & Miron, 1997). Other authors criticize the concept of central bank independence on theoretical grounds, since it lacks democratic justification. As a consequence, concentration on a single objective ignores the population's welfare, specifically in terms of output growth, employment and wealth distribution (see Rochon & Rossi, 2006; Rossi, 2009a). In our context, this debate becomes relevant insofar as if the oil price should be targeted, the central bank would be endowed with an additional instrument of economic and political power. From a democratic point of view this is problematic, since it leaves the central bank with a large autonomy of action, which is required by the nature of the oil price targeting system. However, there are ways to deal with this concern as is already done with the conventional tasks of monetary policy. The US Federal Reserve is obliged to pursue the goals of 'maximum employment, stable prices and moderate long-term interest rate' (Fed, 2014b). In contrast, the European Central Bank (ECB) is instructed to promote merely one objective of first priority, that is, price stability (ECB, 2015). In addition to the existing goals, the central bank may be endowed with a mandate to target the oil price at a path that is determined by the democratic process. This means that it is politics that decides whether or not to allow for this additional objective in general. In analogy to the other monetary policy objectives, it is important to allow the central bank to act within a certain range in order to react to short-run events once the overall objective is given. Since oil price targeting is a long-term proposition, the mandate given to the central bank should not be restricted by daily short-run issues of the political sphere. The central bank may be endowed with it or not. But once this is decided, the monetary authority should be independent of daily political influences. Central banks that are not guaranteed a sufficient space of free action may have difficulties to implement the oil price targeting system.

- *Credibility:* the oil price targeting system may appear as a contrast to today's theoretical framework and practice of central banks. In fact, it can be implemented in broad complementarity rather than in rivalry to the other objectives of monetary policy. However, monetary policy, as is usually practised currently, is based on a supply-determined neoclassical background considering money as (more or less strictly) exogenous. This framework is exclusive in the sense that it tends to focus on a single policy objective, that is, price stability (see ECB, 2015). General equilibrium theory suggests that only prices should be affected by monetary policy. Oil price targeting, in contrast, aims at influencing quantities in the crude oil market as well in the long run. As we argue, this additional goal can be pursued by an additional instrument without impeding the inflation targeting policies. Mutual impacts are given interdependencies since the oil price is part of the general price level and hence affects inflation targeting policies. Yet, we suggest those interdependencies to be limited. All in all, to be credible in targeting the oil price, a central bank has to ground its policy on a theoretical framework that incorporates the endogeneity of money. In order to recognize that monetary policy has the power to affect the real economy, the general equilibrium models have to be abandoned.

- *Economic order:* other questions may arise concerning the economic order of a country. A commodity price that is determined by a political authority may have the appearance of central planning from the perspective of some observers. The accusation usually is that 'targeting an asset price is tantamount to fixing prices, almost certainly to cause misallocations and dislocations that could destabilize asset markets and perhaps the economy' (Sinai, 2009, p. 15). Yet, the oil price targeting system does not reject resource allocation that is driven by market forces. In general, allocation signals of prices trigger complex reactions. Even though market forces without political intervention do not necessarily yield an optimal final outcome, replacing them by political decisions bears considerable risks and difficulties. Finding an optimal price for all goods in the economy is a task that can be traced back to the debates about the transformation problem, which is concerned with the mere existence of such a price system (see, for instance, Baumol, 1974; Meek, 1956; Seton, 1957). However, the case of crude oil is different. We do not claim to have found the optimal price by a scientific proof. Rather, it is found by political considerations. The price should be set so that it gives rise to the optimal quantities of oil production and consumption. In general, this would be hard to find if it was about, for example, a food commodity. We discussed this above. With crude oil, however, there is a simple political purpose or a proposition of environmental science. This reasoning argues for a reduction in oil production and consumption in order to reduce pollution. It is by ecological justification that the oil price should increase in the future and be combined with a tax on crude oil production. All other prices and quantities react correspondingly by adjusting allocation, which is still driven by market forces.

- *Taking account of business cycles:* reducing fossil fuel consumption is an economic, technological and organizational challenge. In a recession, it may be argued that the burden of the increasing oil price is too heavy. In that case, it might raise costs of production inputs, thereby reducing the share of wages in total output, which eventually dampens effective demand. In principle, the central bank may also make the oil price target depend on business cycle conditions. This implies that the oil price is allowed to grow slower, to stagnate or even to drop to a certain degree in times of a recession. On the other hand, keeping the oil price at its target instead of letting it fall in the course of economic stagnation may be a useful tool for monetary policy. When the economy runs the risk of deflation, targeting the oil price at a moderately growing level may counteract this danger. Owing to transparency in the prospected path of the oil price, even expectations of inflation may be lightened up and thus contribute to a normalization of the changes in the general price level.

NOTES

1. There is an intensive debate beyond simple exchange rate pegging that aims at fixing exchange rates institutionally by an international clearing union (see, for instance, Gnos, 2006; Keynes, 1980; Rossi, 2006a, 2015; Wray, 2006). Hence, asset price targeting by monetary policy intervention is incorporated within active research and thus far from being an extraordinary proposition.
2. For an overview of the model modification, see Appendix II.
3. On the other hand, the rising oil price targeted by the central bank affects profits positively.
4. Again, redistributed tax revenues may be used for social security purposes, investment in renewable energy and energy efficiency, and so on.

Conclusion

The global crude oil market is a multilayer issue. On the one hand, it is a well-integrated market and thus may just represent a textbook model. On the other hand, it includes a financial market component that makes it a complex building with various relationships. Disruptions in the crude oil market thus have far-reaching impacts throughout the economy. Likewise, the importance of crude oil as the main fossil energy source makes it a driver of climate change. Understanding the crude oil market is crucial to the implementation of sustainable policies regarding climate change in particular and ecological sustainability in general. The specific interest of this analysis is the impact of monetary policy on the global crude oil market. The insights gained thereby can be used to address the challenges arising from the economic mechanisms identified throughout our investigation.

Allowing monetary policy to have lasting impacts on the economy, by saying that money is endogenous and non-neutral, gives rise to complex dynamics. Monetary policy effects, exerted in an environment that is recognized to be uncertain, open a space for financial speculation to become effective. In this respect we have found the most particular feature of the oil market to be the dual nature of crude oil as a physical commodity and a financial asset. The same good is traded in two different but closely connected markets. The spot market and the futures market aggregate to the crude oil market as a whole. The narrative of the theoretical analysis is the following: expansive monetary policy triggers speculation in the crude oil futures market, which raises the oil price. This improves profit prospects of oil producers, who increase investment expenditures. Sooner or later, rising oil production capacities bring the oil price down to a lower level compared to the beginning.

Overinvestment keeps the oil price low for a longer time, leading to higher oil consumption and therefore a *ceteris paribus* higher oil intensity of the economy. Moreover, speculation in the futures market gives rise to a higher oil price volatility that involves overall financial and economic instability. Theoretical analysis, an SFC model and econometrics taken together confirm this conclusion.

We have outlined a policy proposal that takes both the stability and the ecological sustainability issue into account by combining the currently exist-ing policy ideas. In particular, monetary policy is coordinated with fiscal policy. An oil price target is set and achieved by intervention of the central bank in the futures market along the lines of the use of the strategic petro-leum reserve. As a difference, trading futures contracts, instead of physical oil, allow for more flexibility and better fine-tuning. To prevent imbalances in the spot market, a tax is imposed on oil production. Once the oil price target is achieved, stability in the oil market is guaranteed and speculation is ruled out. By increasing the oil price target step by step to a higher level, oil consumption can be decreased, which is in favour of climate protection. Stability is also established for the rest of the economy, since it becomes clear that investing in oil-consuming equipment will no longer be profitable in the future. Uncertainty with respect to the oil price is removed and it is obvious that investment in renewable energy sources and energy efficiency is sustainable in an ecological as well as economic sense. The insight that the futures and spot market are integrated, so that spot and futures prices move closely together, allows the use of the futures market for economic policy. The proposition at hand makes use of the mechanisms of money and monetary policy that have been examined in this book. Combining the general ecological benefit of a rising oil price with a stable pattern over time commands great effort but it may nevertheless be the most comfortable and efficient way to get out of using fossil energies.

For the future, there is a considerable amount of work to be done. Regarding political intervention in the oil market, institutional details should be enlightened. For instance, be it with the oil price targeting system or with an energy tax in general, the rule of redistribution to the economy so that it does not risk a loss of competition, nor create a moral hazard for oil production, calls for additional research. Moreover, what oil price targeting means for overall inflation has been debated. Yet, how the oil price target has to be set under conditions of general inflation should be investigated closer. While it is not a difficult thing to deflate a price *ex post* by dividing it by a consumer price index, it is more difficult to deter-mine *ex ante* what the oil price ought to be relative to the overall price level. Research in this field may be helpful. More general literature on the macroeconomic effects of oil price shocks has been produced for decades. What is important to know is what the impact that a changing oil price, specifically if the change is lasting, has on the composition of economic output and hence on the structure of employment. Another issue from the macroeconomic perspective is the impact of a rising oil price on income and wealth distribution in the economy. Some groups of the population may be stronger affected than others depending on their consumer basket and on the remuneration scheme of the tax on oil production.

Appendices: SFC model of monetary policy and the crude oil market

APPENDIX I: THE BASIC STRUCTURE

Model Variables

C_d	Consumption goods demand
CG_P	Capital gain of inventories
$C_{oil,d}$	Oil demand
$C_{oil,s}$	Oil supply
C_s	Consumption goods supply (exogenous)
F_L	Net long positions of investors, futures demand
FP^e_I	Expected financial profits of investors
FP_I	Financial profits of investors
FP_P	Financial profits of producers
FPU_I	Undistributed profits of investors
FPU_P	Undistributed financial profits of producers
F_S	Net short positions of producers, futures supply
H_C	Cash held by consumers
H_{CB}	Cash issued by central bank
I_d	Investment demand
IN	Oil inventories
I_s	Investment supply
K	Capital stock
L_B	Loans issued by banks
L_I	Loans granted to investors
L_P	Loans granted to producers
M_B	Maintenance margins at banks
M_I	Maintenance margins of investors
P_B	Profits of banks
P_C	Profits distributed to consumers
P_{CB}	Profits of central bank
$p^e_{I,spot}$	Expected spot price of investors
PP^e_P	Expected profits of producers
p_{fut}	Futures price of oil at the moment of contracting, expiring in the next period
PP_P	Production profits

PPU_P	Undistributed profits of producers
p_{spot}	Spot price of oil
r	Bank lending rate
R_B	Reserves demanded by banks
R_{CB}	Reserves provided by central bank
r_T	Interest rate target (exogenous)
V	Total wealth
V_B	Wealth of banks
V_C	Consumer wealth
V_{CB}	Wealth of central banks
V_I	Wealth of investors
V_P	Wealth of producers
W_d	Wage demand
W_s	Wage supply
Y	Nominal GDP

Model Equations

(4.1) $\quad C_{oil,d} = \delta_0 + \delta_1 {}^*C_s - \delta_2 {}^*p_{spot}$

(4.2) $\quad \Delta IN = \delta_3 {}^*K_{-1} - \gamma {}^*IN_{-1} - C_{oil,s}$

(4.3) $\quad C_{oil,s} = C_{oil,d}$

(4.4) $\quad C_d = C_s$

(4.5) $\quad Y = C_s + p_{spot} {}^*C_{oil,s} + I_s + p_{spot} {}^*\Delta IN$

(4.6) $\quad I_d = \dfrac{\alpha_1 {}^*PP_P^e}{1 + \alpha_2 {}^*L_{P,-1}}$

(4.7) $\quad PP_P^e = PP_{P,-1}$

(4.8) $\quad PP_P = Y - C_d - W_d - r_{-1} {}^*L_{P,-1}$

(4.9) $\quad PPU_P = (1 - s) {}^*PP_P$

(4.10) $\quad I_s = I_d$

(4.11) $\quad \Delta K = I_s$

(4.12) $\quad W_d = \delta_7 {}^*(C_{oil,s} + \Delta IN)$

(4.13) $\quad W_s = W_d$

(4.14) $\quad F_S = F_I$

(4.15) $\quad p_{spot} = p_{fut}$

(4.16) $\quad p_{fut} = \dfrac{\delta_4 + \delta_5 {}^*(F_L + C_{oil,d})}{\delta_6 {}^*K_{-1}}$

(4.17) $\quad F_L = \dfrac{M_I}{m^* p_{fut,-1}}$

(4.18) $\quad \Delta M_I = \dfrac{\beta_0 + \beta_1 {}^* FP_I^e - \beta_2 {}^* r}{1 + \beta_3 {}^* L_I^2}$

(4.19) $\quad \Delta L_1 = \Delta M_1 - FPU_1$

(4.20) $\quad FP_I = (p_{spot} - p_{fut,-1})^* \Delta F_L + \Delta(p_{spot} - p_{fut,-1})^* F_{L,-1} - \dfrac{r_{-1}}{52}^* L_{I,-1}$

(4.21) $\quad FP_I^e = (p_{I,spot}^e - p_{fut,-1})^* \Delta F_{L,-1} + \Delta(p_{I,spot}^e - p_{fut,-1})^* F_{L,-1} - \dfrac{r_{-1}}{52}^* L_{I,-1}$

(4.22) $\quad p_{I,spot}^e = p_{spot,-1} + (p_{spot,-1} - p_{spot,-2})$

(4.23) $\quad FPU_I = FP_I$

(4.24) $\quad FP_P = (p_{fut,-1} - p_{spot})^* \Delta F_s + \Delta(p_{fut,-1} - p_{spot})^* F_{s,-1} + CG_P$

(4.25) $\quad FPU_P = FP_P$

(4.26) $\quad CG_P = \Delta p_{spot}^* IN_{-1}$

(4.27) $\quad \Delta L_P = I_d + \Delta W_s - PPU_P - FPU_P + CG_P + p_{spot}^* \Delta IN$

(4.28) $\quad L_B = L_P + L_I$

(4.29) $\quad P_B = r_{-1}^* L_{B,-1} - r_{T,-1}^* R_{B,-1}$

(4.30) $\quad \Delta R_B = \Delta L_B - \Delta M_B$

(4.31) $\quad M_B = M_I$

(4.32) $\quad P_{CB} = r_{T,-1}^* R_{B,-1}$

(4.33) $\quad r = r_T + D$

(4.34) $\quad P_C = s^* PP_P + P_B + P_{CB}$

(4.35) $\quad \Delta H_C = W_S + P_C - p_{spot}^* C_{oil,d}$

(4.36) $\quad H_{CB} = H_C$

(4.37) $\quad R_{CB} = R_B$

(4.38) $\quad V_P = K + p_{spot}^* IN - L_P$

(4.39) $\quad V_I = M_I - L_I$

(4.40) $\quad V_B = L_B - M_B - R_B$

(4.41) $\quad V_{CB} = R_B - H_C$

(4.42) $\quad V_C = H_C$

(4.43) $\quad V = V_C + V_P + V_I + V_B + V_{CB} = K + p_{spot}^* IN$

Table A.1   Stocks matrix

	Consumers/ households	Producers	Investors	Banks	Central bank	Σ
Capital		$+K$				$+K$
Loans		$-L_P$	$-L_I$	$+L_B$		0
Cash	$+H_C$				$-H_{CB}$	0
Maintenance margin			$+M_I$	$-M_B$		0
Reserves				$-R_B$	$+R_{CB}$	0
Inventories		$+p_{spot}*IN$				$+p_{spot}*IN$
Open interest		$+p_{fut,-1}*F_S$	$-p_{fut,-1}*F_L$			0
Oil balance (due to futures contracting)		$-p_{spot}*F_S$	$+p_{spot}*F_L$			0
Net wealth	$-V_C$	$-V_P$	$-V_I$	$-V_B$	$-V_{CB}$	$-K-p_{spot}*IN$
Σ	0	0	0	0	0	0

Source: Author's elaboration.

Table A.2 *Transactions matrix*

	Consumers/households	Producers		Investors		Banks		Central bank		Σ
		Cu	Cap	Cu	Cap	Cu	Cap	Cu	Cap	
Oil consumption	$-p_{spot}*C_{oil,d}$	$+p_{spot}*C_{oil,s}$								0
Other consumption	$-C_d$	$+C_s$								0
Real investment		$+I_s$	$-I_d$							0
Wages	$+W_s$	$-W_d$								0
Net profits	$+P_C$	$-PP_P$	$+PPU_P$			$-P_B$		$-P_{CB}$		0
Loan interests		$-r_{-1}*L_{P,-1}$		$-r_{t-1}*L_{I,-1}$		$+r_{t-1}*L_{B,-1}$				0
Interests on reserves						$-r_T*R_{B,-1}$		$+r_T*R_{CB,-1}$		0
Stock accumulation		$+p_{spot}*\Delta IN$								$+p_{spot}*\Delta IN_P$
Futures investment/open interest		$+(p_{fut,-1}-p_{spot})*\Delta F_S$ $+\Delta(p_{spot}-p_{fut,-1})*F_{S,-1}$			$+(p_{spot}-p_{fut,-1})*\Delta F_L$ $+\Delta(p_{spot}-p_{fut,-1})*F_{L,-1}$					0
Financial profits		$-FP_P$	$+FPU_P$	$-FP_I$	$+FPU_I$					0

257

Table A.2 (continued)

	Consumers/ households	Producers Cu	Producers Cap	Investors Cu	Investors Cap	Banks Cu	Banks Cap	Central bank Cu	Central bank Cap	Σ
Change in money	$-\Delta H_C$								$+\Delta H_{CB}$	0
Change in loans			$+\Delta L_P$		$+\Delta L_I$		$-\Delta L_B$			0
Change in margin					$-\Delta M_I$		$+\Delta M_B$			0
Change in reserves							$+\Delta R_B$		$-\Delta R_{CB}$	0
Σ	0	0	0	0	0	0	0	0	0	
Stock accumulation		$+\Delta p_{spot}*IN_{-1}$								$+\Delta p_{spot}*IN_{-1}$
Capital accumulation		$+\Delta K = I_s$								$+\Delta K = I_s$

Source: Author's elaboration.

258

APPENDIX II: MODEL EXTENSION WITH POLICY

Additional Variables

F_L^{CB}	Net long positions of central bank
$F_L^{CB,target}$	Central bank target of net long positions (exogenous)
F_L^I	Net long positions of investors (existing before as F_L)
F_L^{tot}	Total long positions
M_{CB}	Maintenance margin of central bank
$p_{fut,target}$	Futures price targeted by central bank (exogenous)
T	Tax on oil production

Modification of Model Equations

(4.8') $$PP_P = Y - C_d - W_d - r_{-1}*L_{P,-1} - T$$

(4.16'') $$F_L^{tot} = \frac{p_{fut}*\delta_6*K_{-1} - \delta_4 - \delta_8*T}{\delta_5} - C_{oil,d}$$

(4.17') $$F_L^I = \frac{M_I}{m*p_{fut,-1}}$$

(4.31') $$M_B = M_I + M_{CB}$$

(6.1) $$p_{fut} = p_{fut,target}$$

(6.2) $$F_L^{CB} = F_L^{tot} - F_L^I$$

(6.3) $$M_{CB} = m*p_{fut,-1}*F_L^{CB}$$

(6.4) $$T = i*(\tau_0 + \tau_1*(p_{fut,target} - 2) + \tau_2*(p_{fut,target} - p_{fut,target,-1})$$
$$+ \tau_3*(F_L^{CB} - F_L^{CB,target}))$$

References

Adams-Kane, J. & Lim, J.J. (2011). Growth Poles and Multipolarity. *The World Bank Development Economics Prospects Group*, Policy Research Working Paper 5712.

Aguilera Klink, F. (1994). Pigou and Coase Reconsidered. *Land Economics, 70*(3), 386–390.

Ahmed, N.M. (2013, May). Die nächste Blase. *Le Monde diplomatique*, pp. 1, 5.

Alesina, A. & Summers, L.H. (1993). Central Bank Independence and Macroeconomic Performance: Some Comparative Evidence. *Journal of Money, Credit and Banking, 25*(2), 151–162.

Alhajji, A.F. & Huettner, D. (2000). OPEC and World Crude Oil Markets from 1973 to 1994: Cartel, Oligopoly, or Competitive? *The Energy Journal, 21*(3), 31–60.

Alquist, R. & Gervais, O. (2011). The Role of Financial Speculation in Driving the Price of Crude Oil. *Bank of Canada Discussion Paper*.

Alquist, R. & Kilian, L. (2010). What Do We Learn From the Price of Crude Oil Futures? *Journal of Applied Econometrics, 25*(4), 539–573.

Ando, A. & Modigliani, F. (1963). The 'Life Cycle' Hypothesis of Saving: Aggregate Implications and Tests. *American Economic Review, 53*(1), 55–84.

Angelopoulou, E. & Gibson, H.D. (2009). The Balance Sheet Channel of Monetary Policy Transmission: Evidence from the United Kingdom. *Economica, 76*(304), 675–703.

Anger, A. & Barker, T. (2015). The Effects of the Financial System and Financial Crises on Global Growth and the Environment. In P. Arestis & M. Sawyer (eds.), *Finance and the Macroeconomics of Environmental Policies*. International Papers in Political Economy Series, London: Palgrave Macmillan, (pp. 153–193).

Anzuini, A., Lombardi, M.J. & Pagano, P. (2013). The Impact of Monetary Policy Shocks on Commodity Prices. *International Journal of Central Banking, 9*(3), 119–144.

Arestis, P. (2015). Coordination of Fiscal with Monetary and Financial Stability Policies Can Better Cure Unemployment. *Review of Keynesian Economics, 3*(2), 233–247.

Arestis, P. & Sawyer, M. (2003). The Nature and Role of Monetary Policy When Money is Endogenous. *Levy Economics Institute of Bard College*, Working Paper No. 374.

Arestis, P. & Sawyer, M. (2004). *Re-Examining Monetary and Fiscal Policy for the 21st Century*. Cheltenham, UK and Northampton, MA, USA: Edward Elgar Publishing.

Arestis, P. & Sawyer, M. (2009). Price and Wage Determination and the Inflation Barrier: Moving Beyond the Phillips Curve. In C. Gnos & L.-P. Rochon (eds.), *Monetary Policy and Financial Stability: A Post-Keynesian Agenda*. Cheltenham, UK and Northampton, MA, USA: Edward Elgar Publishing, (pp. 32–47).

Argus. (2015). *Argus Crude: daily reporting on the global crude markets*. Accessed 26 November 2015 at http://www.argusmedia.com/Crude-Oil/Argus-Crude.

Arora, V. & Tanner, M. (2013). Do Oil Prices Respond to Real Interest Rates? *Energy Economics, 36*, 546–555.

Asensio, A. (2007). Monetary and Budgetary-Fiscal Policy Interactions in a Keynesian Context: Revisiting Macroeconomic Governance. In P. Arestis, E. Hein & E. Le Heron (eds.), *Aspects of Modern Monetary and Macroeconomic Policies*. London: Palgrave Macmillan, (pp. 80–105).

Asensio, A., Lang, D. & Charles, S. (2010). Post-Keynesian Modelling: Where Are We, and Where Are We Going To? *Munich Personal RePEc Archive Paper*, No. 30726.

Associated Press. (2016). US auto sales hit an all-time high in 2015. *The Denver Post*. Accessed 19 January 2016 at http://www.denverpost.com/2016/10/04/new-car-sales-prices-deals/.

Atesoglu, H.S. (2003–04). Monetary Transmission: Federal Funds Rate and Prime Rate. *Journal of Post Keynesian Economics, 26*(2), 357–362.

Aysun, U. & Hepp, R. (2013). Identifying the Balance Sheet and the Lending Channels of Monetary Transmission: A Loan-Level Analysis. *Journal of Banking & Finance, 37*(8), 2812–2822.

Baek, J. & Koo, W.W. (2014). On the Upsurge of U.S. Food Prices Revisited. *Economic Modelling, 42*, 272–276.

Ball, L., Mankiw, N.G., Romer, D., Akerlof, G.A., Rose, A., Yellen, J. et al. (1988). The New Keynesian Economics and the Output-Inflation Trade-Off. *Brookings Papers on Economic Activity, 19*(1), 1–82.

Bandyopadhyay, K.R. (2008). OPEC's Price-Making Power. *Economic and Political Weekly, 43*(46), 18–21.

Bandyopadhyay, K.R. (2009). Does OPEC Act as a Residual Producer? *Munich Personal RePEc Archive Paper*, No. 25841.

Bank for International Settlements [BIS]. (2013). OTC derivatives statistics at end-June 2013. *Bank for International Settlements, Statistical release*, November 2013.

Barnes, P. (2008, 1 December). Cap and Dividend, Not Trade: Making Polluters Pay. *Scientific American*. Accessed 7 January 2016 at http://www.scientificamerican.com/article/cap-and-divident-not-trade/.

Barro, R.J. & Gordon, D.B. (1983). A Positive Theory of Monetary Policy in a Natural Rate Model. *Journal of Political Economy, 91*(4), 589–610.

Barro, R.J. & Grossman, H.I. (1974). Suppressed Inflation and the Supply Multiplier. *Review of Economic Studies, 41*(1), 87–104.

Barsky, R.B. & Kilian, L. (2004). Oil and the Macroeconomy Since the 1970s. *Journal of Economic Perspectives, 18*(4), 115–134.

Basistha, A. & Kurov, A. (2015). The Impact of Monetary Policy Surprises on Energy Prices. *Journal of Futures Markets, 35*(1), 87–103.

Baumol, W.J. (1972). On Taxation and the Control of Externalities. *American Economic Review, 62*(3), 307–322.

Baumol, W.J. (1974). The Transformation of Values: What Marx 'Really' Meant (An Interpretation). *Journal of Economic Literature, 12*(1), 51–62.

Bayer, P., Dolan, L. & Urpelainen, J. (2013). Global Patterns of Renewable Energy Innovation, 1990–2009. *Energy for Sustainable Development, 17*(3), 288–295.

Bénicourt, E. & Guerrien, B. (2008). *La théorie économique néoclassique: microéconomie, macroéconomie et théorie des jeux*. Paris: La Découverte.

Berck, P. & Cecchetti, S.G. (1985). Portfolio Diversification, Futures Markets, and Uncertain Consumption Prices. *American Journal of Agricultural Economics, 67*(3), 497–507.

Bernanke, B.S. & Gertler, M. (1995). Inside the Black Box: The Credit Channel of Monetary Policy Transmission. *Journal of Economic Perspectives, 9*(4), 27–48.

Bernanke, B.S. & Gertler, M. (2001). Should Central Banks Respond to Movements in Asset Prices? *American Economic Review, 91*(2), 253–257.

Bernanke, B.S., Gertler, M. & Gilchrist, S. (1996). The Financial Accelerator and the Flight to Quality. *Review of Economics and Statistics, 78*(1), 1–15.

Bernanke, B.S., Gertler, M. & Watson, M. (1997). Systematic Monetary Policy and the Effects of Oil Price Shocks. *Brookings Papers on Economic Activity, 28*(1), 91–142.

Bernanke, B.S. & Mihov, I. (1995). Measuring Monetary Policy. *National Bureau of Economic Research Working Paper*, No. 5145.

Bernanke, B.S. & Mishkin, F.S. (1997). Inflation Targeting: A New Framework for Monetary Policy? *Journal of Economic Perspectives, 11*(2), 97–116.

Bernier, A. (2007, December). Monopoly mit dem Weltklima. *Le Monde diplomatique*, pp. 12–13.

Bernow, S., Costanza, R., Daly, H., DeGennaro, R., Erlandson, D., Ferris, D. et al. (1998). Society News: Ecological Tax Reform. *BioScience, 48*(3), 193–196.

Bibow, J. (2006). Liquidity Preference Theory. In P. Arestis & M. Sawyer (eds.), *Handbook of Alternative Monetary Economics*. Cheltenham, UK and Northampton, MA, USA: Edward Elgar Publishing, (pp. 328–345).

Bickley, J.M. (2012). The Federal Excise Tax on Gasoline and the Highway Trust Fund: A Short History. *Congressional Research Service*, RL30304.

Bikhchandani, S. & Sharma, S. (2001). Herd Behavior in Financial Markets. *International Monetary Fund Staff Papers, 47*(3), 279–310.

Blanchard, O.J. & Galí, J. (2007). The Macroeconomic Effects of Oil Shocks: Why Are the 2000s so Different from the 1970s? *National Bureau of Economic Research Working Paper*, No. 13368.

Blanchard, O.J. & Illing, G. (2006). *Makroökonomie – 4., aktualisierte und erweiterte Auflage*. München: Pearson Studium.

Board of Governors of the Federal Reserve System (US). (2015a). *Capacity Utilization: Oil and gas extraction* [CAPUTLG211S]. Accessed 4 June 2015 at https://research.stlouisfed.org/fred2/series/CAPUTLG211S/.

Board of Governors of the Federal Reserve System (US). (2015b). *Industrial Capacity: Mining: Oil and gas extraction* [CAPG211SQ]. Accessed 10 June 2015 at https://research.stlouisfed.org/fred2/series/CAPG211SQ/.

Board of Governors of the Federal Reserve System (US). (2015c). *Industrial Production: Mining: Drilling oil and gas wells* [IPN213111N]. Accessed 4 June 2015 at https://research.stlouisfed.org/fred2/series/IPN213111N/.

Board of Governors of the Federal Reserve System (US). (2015d). *Trade Weighted US Dollar Index: Broad* [DTWEXB]. Accessed 11 April 2015 at http://research.stlouisfed.org/fred2/series/DTWEXB.

Bodenstein, M., Guerrieri, L. & Kilian, L. (2012). Monetary Policy Responses to Oil Price Fluctuations. *IMF Economic Review, 60*(4), 470–504.

Bordo, M.D., Dueker, M.J. & Wheelock, D.C. (2007). Monetary Policy and Stock Market Booms and Busts in the 20th Century. *Federal Reserve Bank of St. Louis Working Paper*, No. 2007-020A.

Borio, C. & Disyatat, P. (2009). Unconventional Monetary Policies: An Appraisal. *Bank for International Settlements Working Paper*, No. 292.

Borio, C., McCauley, R. & McGuire, P. (2011). Global Credit and Domestic Credit Booms. *Bank for International Settlements Quarterly Review*, September 2011.

Boyce, J.K. & Riddle, M. (2007). Cap and Dividend: How to Curb Global Warming While Protecting the Incomes of American Families. *University of Massachusetts Amherst Political Economy Research Institute Working Paper,* No. 150.

Brémond, V., Hache, E. & Mignon, V. (2012). Does OPEC Still Exist as a Cartel? An Empirical Investigation. *Energy Economics, 34*(1), 125–131.

Brigida, M. (2014). The Switching Relationship Between Natural Gas and Crude Oil Prices. *Energy Economics, 43*, 48–55.

British Petroleum [BP]. (2015). *BP Statistical Review of World Energy June 2015.* London.

Brittan, S. (2009). Should, or Can, Central Banks Target Asset Prices? A Symposium of Views. *The International Economy*, Fall, 9.

Brunner, K., Fratianni, M., Jordan, J.L., Meltzer, A.H. & Neumann, M.J.M. (1973). Fiscal and Monetary Policies in Moderate Inflation: Case Studies in Three Countries. *Journal of Money, Credit and Banking, 5*(1), 313–353.

Büyükşahin, B. & Harris, J.H. (2011). Do Speculators Drive Crude Oil Prices? *The Energy Journal, 32*(2), 167–202.

Büyükşahin, B. & Robe, M.A. (2011). Does 'Paper Oil' Matter? Energy Markets' Financialization and Equity-Commodity Co-Movements. *CFTC and IEA Working Paper*, Washington, DC and Paris.

Campillo, M. & Miron, J. (1997). Why Does Inflation Differ Across Countries? In C.D. Romer & D.H. Romer (eds.), *Reducing Inflation: Motivation and Strategy.* Chicago: University of Chicago Press, (pp. 335–357).

Canuto, O. & Cavallari, M. (2013). Asset Prices, Macroprudential Regulation, and Monetary Policy. *The World Bank Economic Premise*, No 116.

Carbon Trade Watch. (2013). *It is time to scrap the ETS!* Accessed 11 September 2015 at http://www.carbontradewatch.org/articles/time-to-scrap-the-ets.html.

Cardim de Carvalho, F.J. (2013). Keynes and the Endogeneity of Money. *Review of Keynesian Economics, 1*(4), 431–446.

Carlstrom, C.T. & Fuerst, T.S. (2005). Oil Prices, Monetary Policy, and the Macroeconomy. *Federal Reserve Bank of Cleveland Policy Discussion Paper*, No. 10.

Caruana, J. (2013). *Ebbing global liquidity and monetary policy interactions.* Speech delivered at the Central Bank of Chile Fifth Summit Meeting of Central Banks on Inflation Targeting: Global liquidity, capital flows and policy coordination. Santiago, Chile, 15 November.

Cavallo, M. (2008). Oil Prices and Inflation. *Federal Reserve Bank of San Francisco Economic Letter*, No. 31.

Cecchetti, S.G. & Moessner, R. (2008). Commodity Prices and Inflation Dynamics. *Bank for International Settlements Quarterly Review*, December 2008.

Cencini, A. (2000). World Monetary Disorders: Exchange Rate Erratic Fluctuations. *Centro di Studi Bancari Quaderni di Ricerca*, No. 2.

Cencini, A. (2003a). IS-LM: A Final Rejection. In L.-P. Rochon & S. Rossi (eds.), *Modern Theories of Money: The Nature and Role of Money in Capitalist Economies*. Cheltenham, UK and Northampton, MA, USA: Edward Elgar Publishing, (pp. 295–321).

Cencini, A. (2003b). Neoclassical, New Classical and New Business Cycle Economics: A Critical Survey. *Centro di Studi Bancari Quaderni di Ricerca*, No. 10.

Cencini, A. (2005). *Macroeconomic Foundations of Macroeconomics*. London: Routledge.

Cencini, A. (2012). Towards a Macroeconomic Approach to Macroeconomics. In C. Gnos & S. Rossi (eds.), *Modern Monetary Macroeconomics: A New Paradigm for Economic Policy*. Cheltenham, UK and Northampton, MA, USA: Edward Elgar Publishing, (pp. 39–68).

Cevik, S. & Teksoz, K. (2012). Lost in Transmission? The Effectiveness of Monetary Policy Transmission Channels in the GCC Countries. *International Monetary Fund Working Paper*, No. 12/191.

Chant, J. (2003). Financial Stability as a Policy Goal. *Bank of Canada Essays on Financial Stability*, Technical Report No. 95.

Chen, S.-S. (2009). Oil Price Pass-through into Inflation. *Energy Economics, 31*(1), 126–133.

Chiu, P. (2003). Transparency versus Constructive Ambiguity in Foreign Exchange Intervention. *Bank for International Settlements Working Paper*, No. 144.

Cifarelli, G. & Paladino, G. (2010). Oil Price Dynamics and Speculation – A Multivariate Financial Approach. *Energy Economics, 32*(2), 363–372.

Claessens, S. & Ratnovski, L. (2014). What is Shadow Banking? *International Monetary Fund Working Paper*, No. 14/25.

Clower, R.W. (1967). A Reconsideration of the Microfoundations of Monetary Theory. *Western Economic Journal, 6*(1), 1–8.

Coase, R.H. (1960). The Problem of Social Cost. *Journal of Law and Economics, 3*, 1–44.

Colander, D. (1996). AS/AD, AE/AP, IS/LM and Z. In P. Arestis (ed.), *Keynes, Money and the Open Economy: Essays in Honour of Paul Davidson: Volume One*. Cheltenham, UK and Brookfield, VT, USA: Edward Elgar Publishing, (pp. 22–33).

Colander, D. (2001). Effective Supply and Effective Demand. *Journal of Post Keynesian Economics, 23*(3), 375–381.

Committee on Climate Change. (2009). *Meeting Carbon Budgets – the need for a step change.* Accessed 11 September 2015 at https://www.theccc.org.uk/publication/meeting-carbon-budgets-the-need-for-a-step-change-1st-progress-report/.

Commodity Futures Trading Commission [CFTC]. (2014). *Commitments of Traders: Historical compressed.* Accessed 21 October 2014 at http://www.cftc.gov/MarketReports/CommitmentsofTraders/HistoricalCompressed/index.htm.

Commodity Futures Trading Commission [CFTC]. (2015). *CFTC Glossary.* Accessed 11 April 2015 at http://www.cftc.gov/ConsumerProtection/EducationCenter/CFTCGlossary/index.htm#C.

Considine, T.J. (2006). Is the Strategic Petroleum Reserve Our Ace in the Hole? *The Energy Journal, 27*(3), 91–112.

Cooper, J.C.B. (2003). Price Elasticity of Demand for Crude Oil: Estimates for 23 Countries. *OPEC Review, 27*, 1–8.

Cottrell, A. (1994). Post Keynesian Monetary Economics: A Critical Survey. *Cambridge Journal of Economics, 18*(6), 587–605.

Creel, J., Hubert, P. & Viennot, M. (2013). Assessing the Interest Rate and Bank Lending Channels of ECB Monetary Policies. *L'Observatoire français des conjonctures économiques Working Paper*, No. 2013–25.

Cukierman, A., Webb, S.B. & Neyapti, B. (1992). Measuring the Independence of Central Banks and Its Effect on Policy Outcomes. *The World Bank Economic Review, 6*(3), 353–398.

Dalziel, P. (1999–2000). A Post Keynesian Theory of Asset Price Inflation with Endogenous Money. *Journal of Post Keynesian Economics, 22*(2), 227–245.

D'Amico, S., English, W., López-Salido, D. & Nelson, E. (2012). The Federal Reserve's Large-Scale Asset Purchase Programmes: Rationale and Effects. *The Economic Journal, 564*, F415–F446.

Davidson, P. (2002). *Financial Markets, Money and the Real World.* Cheltenham, UK and Northampton, MA, USA: Edward Elgar Publishing.

Davidson, P. (2006). Exogenous versus Endogenous Money: The Conceptual Foundations. In M. Setterfield (ed.), *Complexity, Endogenous Money and Macroeconomic Theory.* Cheltenham, UK and Northampton, MA, USA: Edward Elgar Publishing, (pp. 141–149).

Davidson, P. (2008). Crude Oil Prices: 'Market Fundamentals' or Speculation? *Challenge, 51*(4), 110–118.

Davidson, P. & Weintraub, S. (1973). Money as Cause and Effect. *The Economic Journal, 83*(332), 1117–1132.

De Long, J.B., Shleifer, A., Summers, L.H. & Waldmann, R.J. (1990). Positive Feedback Investment Strategies and Destabilizing Rational Speculation. *Journal of Finance, 45*(2), 379–395.

Dow, S.C. (2006). Endogenous Money: Structuralist. In P. Arestis & M. Sawyer (eds.), *Handbook of Alternative Monetary Economics*. Cheltenham, UK and Northampton, MA, USA: Edward Elgar Publishing, (pp. 35–51).

Dvir, E. & Rogoff, K. (2014). Demand Effects and Speculation in Oil Markets: Theory and Evidence. *Journal of International Money and Finance, 42*, 113–128.

Ehrmann, M. & Fratzscher, M. (2004). Taking Stock: Monetary Policy Transmission to Equity Markets. *Journal of Money, Credit and Banking, 36*(4), 719–737.

Eichenbaum, M. (1992). Comment on 'Interpreting the Macroeconomic Time Series Facts: The Effects of Monetary Policy' by C.A. Sims. *European Economic Review, 36*(5), 1001–1011.

Eichenbaum, M. and Evans, C.L. (1995). Some Empirical Evidence on the Effects of Shocks to Monetary Policy on Exchange Rates. *Quarterly Journal of Economics, 110*(4), 975–1009.

Enders, W. (2014). *Applied Econometric Time Series* (4th ed.). New York: John Wiley & Sons, Inc.

Energy Information Administration [EIA]. (2013). *Spread narrows between Brent and WTI crude oil benchmark prices*. Accessed 8 January 2015 at http://www.eia.gov/todayinenergy/detail.cfm?id=12391.

Energy Information Administration [EIA]. (2015a). *How much carbon dioxide is produced when different fuels are burned?* Accessed 29 October 2015 at http://www.eia.gov/tools/faqs/faq.cfm?id=73&t=11.

Energy Information Administration [EIA]. (2015b). *International energy statistics*. Accessed 4 June 2015 at www.eia.gov/outlooks/steo/report/global_oil.cfm.

Energy Information Administration [EIA]. (2015c). *Petroleum and other liquids*. Accessed 8 January 2015 at http://www.eia.gov/petroleum/data.cfm#prices.

Energy Information Administration [EIA]. (2015d). *Petroleum and other statistics: Total stocks*. Accessed 9 June 2015 at https://www.eia.gov/petroleum/data.cfm.

Engel, C. (2010). Exchange Rate Policies. *Bank for International Settlements Papers*, No 52.

Ennis, M.H. & Keister, T. (2008). Understanding Monetary Policy Implementation. *Federal Reserve Bank of Richmond Economic Quarterly, 94*(3), 235–263.

Epstein, G.A. (2005). Introduction: Financialization and the World

Economy. In G.A. Epstein (ed.), *Financialization and the World Economy*. Cheltenham, UK and Northampton, MA, USA: Edward Elgar Publishing, (pp. 3–16).

Erdös, P. (2012). Have Oil and Gas Prices Got Separated? *Energy Policy, 49*, 707–718.

Erturk, K.A. (2006). Speculation, Liquidity Preference and Monetary Circulation. In P. Arestis & M. Sawyer (eds.), *Handbook of Alternative Monetary Economics*. Cheltenham, UK and Northampton, MA, USA: Edward Elgar Publishing, (pp. 455–470).

Estrella, A. (2002). Securitization and the Efficacy of Monetary Policy. *Federal Reserve Bank of New York Economic Policy Review, 8*(1), 243–255.

European Central Bank [ECB]. (2015). *Objective of monetary policy*. Accessed 22 October 2015 at https://www.ecb.europa.eu/mopo/intro/objective/html/in dex.en.html.

European Commission. (2015). *The EU Emissions Trading System (EU ETS)*. Accessed 8 September 2015 at https://ec.europa.eu/clima/policies/ets_en.

Fama, E.F. (1970). Efficient Capital Markets: A Review of Theory and Empirical Work. *Journal of Finance, 25*(2), 383–417.

Fama, E.F. (1991). Efficient Capital Markets: II. *Journal of Finance, 46*(5), 1575–1617.

Fan, Y. & Xu, J. (2011). What Has Driven Oil Prices Since 2000? A Structural Change Perspective. *Energy Economics, 33*(6), 1082–1094.

Fattouh, B. (2010). Oil Market Dynamics Through the Lens of the 2002–2009 Price Cycle. *Oxford Institute for Energy Studies*, WPM 39.

Fattouh, B. (2011). An Anatomy of the Crude Oil Pricing System. *Oxford Institute for Energy Studies*, WPM 40.

Fattouh, B. & Mahadeva, L. (2012). Assessing the Financialization Hypothesis. *Oxford Institute for Energy Studies*, WPM 49.

Fattouh, B., Kilian, L. & Mahadeva, L. (2012). The Role of Speculation in Oil Markets: What Have We Learned So Far? *Oxford Institute for Energy Studies*, WPM 45.

Federal Reserve Bank of New York. (2015). *System Open Market Account Holdings*. Accessed 10 April 2015 at http://www.newyorkfed.org/markets/soma/sys open_accholdings.html.

Federal Reserve System [Fed]. (2006). *Press Release: For immediate release*. Accessed 11 May 2015 at https://www.federalreserve.gov/news-events/press/monetary/20060629a.htm.

Federal Reserve System [Fed]. (2012). *Press Release: For immediate release*. Accessed 14 April 2015 at http://www.federalreserve.gov/news-events/press/monetary/20120913a.htm.

Federal Reserve System [Fed]. (2014a). *Monetary policy releases.* Accessed 14 April 2015 at http://www.federalreserve.gov/newsevents/press/mone tary/2015monetary.htm.

Federal Reserve System [Fed]. (2014b). *Statement on longer-run goals and monetary policy strategy.* Accessed 17 December 2014 at http://www. federalreserve.gov/monetarypolicy/default.htm.

Federal Reserve System [Fed]. (2015). *Selected interest rates (daily) – H.15: Historical data.* Accessed 11 April 2015 at http://www.federalreserve.gov/releases/h15/data.htm.

Fernández de Córdoba, G. & Kehoe, T.J. (2009). The Current Financial Crisis: What Should We Learn from the Great Depressions of the Twentieth Century? *Federal Reserve Bank of Minneapolis Research Department Staff Report,* No. 421.

Filardo, A., Ma, G. & Mihaljek, D. (2011). Exchange Rates and Monetary Policy Frameworks in EMEs. *Bank for International Settlements Papers,* No. 57.

Financial Stability Board [FSB] (2013). *Global Shadow Banking Monitoring Report 2013.* Accessed 21 November 2014 at http://www.financialstabilityboard.org/2013/11/r_131114/.

Forder, J. (2006). Monetary Policy. In P. Arestis & M. Sawyer (eds.), *Handbook of Alternative Monetary Economics.* Cheltenham, UK and Northampton, MA, USA: Edward Elgar Publishing, (pp. 224–241).

Foundation for the Economics of Sustainability [Feasta]. (2008). *Cap & share: A fair way to cut greenhouse emissions.* Accessed 7 January 2016 at http://www.capglobalcarbon.org/the-thinking-behind-capglobalcarbon/.

Frank, R.H. (2008). *Microeconomics and Behavior – Seventh Edition.* New York: McGraw-Hill.

Frankel, J.A. (1984). Commodity Prices and Money: Lessons from International Finance. *American Journal of Agricultural Economics, 66*(5), 560–566.

Frankel, J.A. (2006). The Effect of Monetary Policy on Real Commodity Prices. *National Bureau of Economic Research Working Paper,* No. 12713.

Frankel, J.A. (2014). Effects of Speculation and Interest Rates in a 'Carry Trade' Model of Commodity Prices. *Journal of International Money and Finance, 42,* 88–112.

Frankel, J.A. & Rose, A.K. (2010). Determinants of Agricultural and Mineral Commodity Prices. *Harvard University Research Working Paper,* No. RWP10-038.

Friedman, M. (1956). The Quantity Theory of Money: A Restatement. In *Studies in the Quantity Theory of Money.* Chicago: University of Chicago Press, (pp. 3–21).

Friedman, M. (1959). The Demand for Money: Some Theoretical and Empirical Results. *American Economic Review, 49*(2), 525–527.

Friedman, M. (1968). The Role of Monetary Policy. *American Economic Review, 58*(1), 1–12.

Friedman, M. (1986). The Resource Cost of Irredeemable Paper Money. *Journal of Political Economy, 94*(3), 642–647.

Gagnon, J., Raskin, M., Remache, J. & Sack, B. (2011). The Financial Market Effects of the Federal Reserve's Large-Scale Asset Purchases. *International Journal of Central Banking, 7*(1), 3–43.

Gambacorta, L. & Marques-Ibanez, D. (2011). The Bank Lending Channel: Lessons from the Crisis. *Bank for International Settlements Working Paper*, No. 345.

Gilbertson, T. & Reyes, O. (2009). Carbon Trading: How it Works and Why it Fails. *Critical currents – Dag Hammarskjöld Foundation Occasional Paper*, No. 7.

Gilchrist, S. & Zakrajsek, E. (2013). The Impact of the Federal Reserve's Large-Scale Asset Purchase Programs on Corporate Credit Risk. *Finance and Economics Discussion Series, Federal Reserve Board*, Washington, DC.

Glick, R. & Leduc, S. (2011). Are Large-Scale Asset Purchases Fueling the Rise in Commodity Prices? *Federal Reserve Board of San Francisco Economic Letter*, No. 2011–10.

Global Justice Now. (2015). *Food speculation*. Accessed 28 August 2015 at http://www.globaljustice.org.uk/food-speculation.

Gnos, C. (2003). Circuit Theory as an Explanation of the Complex Real World. In L.-P. Rochon & S. Rossi (eds.), *Modern Theories of Money: The Nature and Role of Money in Capitalist Economies*. Cheltenham, UK and Northampton, MA, USA: Edward Elgar Publishing, (pp. 322–338).

Gnos, C. (2006). Reforming the International Payment System: An Assessment. In L.-P. Rochon & S. Rossi (eds.), *Monetary and Exchange Rate Systems: A Global View of Financial Crises*. Cheltenham, UK and Northampton, MA, USA: Edward Elgar Publishing, (pp. 127–139).

Gnos, C. (2007). Monetary Policy from a Circuitist Perspective. In P. Arestis, E. Hein & E. Le Heron (eds.), *Aspects of Modern Monetary and Macroeconomic Policies*. London: Palgrave Macmillan, (pp. 106–122).

Gnos, C. & Rochon, L.-P. (2007). The New Consensus and Post-Keynesian Interest Rate Policy. *Review of Political Economy, 19*(3), 369–386.

Godley, W. & Lavoie, M. (2012). *Monetary Economics: An Integrated Approach to Credit, Money, Income, Production and Wealth* (2nd ed.). Chippenham and Eastbourne: Palgrave Macmillan.

Goodfriend, M. (2002). Interest on Reserves and Monetary Policy. *Federal Reserve Bank of New York Economic Policy Review, 8*(1), 1–8.

Goodfriend, M. (2007). How the World Achieved Consensus on Monetary Policy. *National Bureau of Economic Research Working Paper*, No. 13580.

Goodfriend, M. & King, R.G. (1997). The New Neoclassical Synthesis and the Role of Monetary Policy. In B.S. Bernanke & J.J. Rotemberg (eds.), *NBER Macroeconomics Annual 1997*. Cambridge, MA: MIT Press, (pp. 231–283).

Gorton, G. & Rouwenhorst, K.G. (2006). Facts and Fantasies about Commodity Futures. *Financial Analyst Journal: Select Financial Analyst Journal Author Summaries, 62*(2), 73–75.

Granger, C.W. (2004). Time Series Analysis, Cointegration, and Applications. *American Economic Review, 94*(3), 421–425.

Graziani, A. (2003). *The Monetary Theory of Production*. Cambridge: Cambridge University Press.

Greenberger, M. (2010). Out of the Black Hole: Regulatory Reform of the Over-the-Counter Derivatives Market. *University of Maryland Legal Studies Research Paper*, No. 2010–51.

Greenwald, B. & Stiglitz, J.E. (1987). Keynesian, New Keynesian and New Classical Economics. *Oxford Economic Papers, 39*(1), 119–132.

Greenwood, J., Hercowitz, Z. & Huffman, G.W. (1988). Investment, Capacity Utilization and the Real Business Cycle. *American Economic Review, 78*(3), 402–417.

Griffin, J.M. & Vielhaber, L.M. (1994). OPEC Production: The Missing Link. *The Energy Journal, 15*, Special Issue on the Changing World Petroleum Market, 115–132.

Grossman, S.J. & Stiglitz, J.E. (1980). On the Impossibility of Informationally Efficient Markets. *American Economic Review, 70*(3), 393–408.

Guerrien, B. & Gun, O. (2011). Efficient Market Hypothesis: What Are We Talking About? *Real-World Economics Review, 56*, 19–30.

Hamburg Institute of International Economics [HWWI]. (2015). *Coal USD Price Index*. Datastream.

Hamilton, J.D. (1994). *Time Series Analysis*. Princeton: Princeton University Press.

Hamilton, J.D. (2009). Causes and Consequences of the Oil Shock of 2007–08. *Brookings Papers on Economic Activity, 40*(1), 215–261.

Hamilton, J.D. (2012). Oil Prices, Exhaustible Resources, and Economic Growth. *National Bureau of Economic Research Working Paper*, No. 17759.

Hartwig, J. (2006). Explaining the Aggregate Price Level with Keynes's

Principle of Effective Demand. *Review of Social Economy, 64*(4), 469–492.

Hein, E. & Truger, A. (2011). Finance-Dominated Capitalism in Crisis – The Case for a Keynesian New Deal at the European and the Global Level. In P. Arestis & M. Sawyer (eds.), *New Economics as Mainstream Economics*, International Papers in Political Economy Series. London: Palgrave Macmillan, (pp. 190–230).

Hendry, D.F. (2004). The Nobel Memorial Prize for Clive W.J. Granger. *The Scandinavian Journal of Economics, 106*(2), 187–213.

Hodson, P.V. (2013). History of Environmental Contamination by Oil Sands Extraction. *Proceedings of the National Academy of Sciences, 110*(5), 1569–1570.

Hoover, K.D. (2008). Causality in Economics and Econometrics. In *The New Palgrave Dictionary of Economics* (2nd ed.). Accessed 9 April 2015 at http://www.dictionaryofeconomics.com/article?id=pde2008_C000569&edition=current&q=causality%20in%20economics&topicid=&result_number=1.

Hotelling, H. (1931). The Economics of Exhaustible Resources. *Journal of Political Economy, 39*(2), 137–175.

Howells, P. (1995). The Demand for Endogenous Money. *Journal of Post Keynesian Economics, 18*(1), 89–106.

Howells, P. (2006). The Endogeneity of Money: Empirical Evidence. In P. Arestis & M. Sawyer (eds.), *Handbook of Alternative Monetary Economics*. Cheltenham, UK and Northampton, MA, USA: Edward Elgar Publishing, (pp. 52–68).

Interagency Task Force on Commodity Markets [ITF]. (2008). *Interim Report on Crude Oil*. Washington, DC.

International Energy Agency [IEA]. (2008). *World Energy Outlook 2008*. Paris.

International Energy Agency [IEA]. (2012). *World Energy Outlook 2012*. Paris.

International Energy Agency [IEA]. (2013). *World Energy Outlook 2013 – Executive Summary*. Paris.

Investopedia. (2015). *Market operations – Margin requirements*. Accessed 10 March 2015 at http://www.investopedia.com/terms/m/margin.asp.

Irwin, S.H., Sanders, D.R. & Merrin, R.P. (2009). Devil or Angel? The Role of Speculation in the Recent Commodity Price Boom (and Bust). *Journal of Agricultural and Applied Economics, 41*(2), 377–391.

Jehle, G.A. & Reny, P.J. (2011). *Advanced Microeconomic Theory* (3rd ed.). Harlow: Pearson Education Limited.

Johansen, S. (1991). Estimation and Hypothesis Testing of Cointegration

Vectors in Gaussian Vector Autoregressive Models. *Econometrica, 59*(6), 1551–1580.

Jones, D.M. (1999). Fed Policy, Financial Market Efficiency, and Capital Flows. *Journal of Finance, 54*(4), 1501–1507.

Jones, S.L. & Netter, J.M. (2008). Efficient Capital Markets. In *The Concise Encyclopedia of Economics*. Accessed 17 September 2014 at http://www.econlib.org/library/Enc/EfficientCapitalMarkets.html.

Joyce, M., Miles, D., Scott, A. & Vayanos, D. (2012). Quantitative Easing and Unconventional Monetary Policy – An Introduction. *The Economic Journal, 564*, F271–F288.

Kaldor, N. (1939). Speculation and Economic Stability. *Review of Economic Studies, 7*(1), 1–27.

Kaldor, N. (1970). The New Monetarism. *Lloyds Bank Review, 97*, 1–18.

Kaldor, N. (1982). *The Scourge of Monetarism*. New York: Oxford University Press.

Kaldor, N. (1985). How Monetarism Failed. *Challenge, 28*(2), 4–13.

Kalecki, M. (1987). *Krise und Prosperität im Kapitalismus: Ausgewählte Essays 1933–1971*. K. Laski & J. Pöschl (Hrsg.), Marburg: Metropolis.

Kapetanios, G., Mumtaz, H., Stevens, I. & Theodoridis, K. (2012). Assessing the Economy-Wide Effects of Quantitative Easing. *The Economic Journal, 564*, F316–F347.

Kashyap, A.K. & Stein, J.C. (2000). What Do a Million Observations on Banks Say About the Transmission of Monetary Policy? *American Economic Review, 90*(3), 407–428.

Kaufmann, R.K. (2011). The Role of Market Fundamentals and Speculation in Recent Price Changes for Crude Oil. *Energy Policy, 39*(1), 105–115.

Kaufmann, R.K. & Ullman, B. (2009). Oil Prices, Speculation, and Fundamentals: Interpreting Causal Relations among Spot and Futures Prices. *Energy Economics, 31*(4), 550–558.

Kaufmann, R.K., Dees, S., Karadeloglou, P. & Sánchez, M. (2004). Does OPEC Matter? An Econometric Analysis of Oil Prices. *The Energy Journal, 25*(4), 67–90.

Kehoe, T.J. & Prescott, E.C. (2007). Great Depressions of the Twentieth Century. In T.J. Kehoe & E.C. Prescott (eds.), *Great Depressions of the Twentieth Century*. Minneapolis: Federal Reserve Bank of Minneapolis, (pp. 1–23).

Keynes, J.M. (1930a/2011). *A Treatise on Money, Volume I: The Pure Theory of Money*. New York: Harcourt, Brace and Company.

Keynes, J.M. (1930b/2011). *A Treatise on Money, Volume II: The Applied Theory of Money*. New York: Harcourt, Brace and Company.

Keynes, J.M. (1936/1997). *The General Theory of Employment, Interest, and Money*. New York: Prometheus Books.

Keynes, J.M. (1937). The General Theory of Employment. *Quarterly Journal of Economics, 51*(2), 209–223.

Keynes, J.M. (1980). *The Collected Writings of John Maynard Keynes* (vol. XXV *Activities 1940–1944. Shaping the Post-War World: The Clearing Union*). London and Basingstoke: Macmillan.

Kilian, L. (2009a). Did Unexpectedly Strong Economic Growth Cause the Oil Price Shock of 2003–2008? *Centre for Economic Policy Research Discussion Paper*, No. 7265.

Kilian, L. (2009b). Not All Oil Price Shocks Are Alike: Disentangling Demand and Supply Shocks in the Crude Oil Market. *American Economic Review, 99*(3), 1053–1069.

Kilian, L. (2010a). Oil Price Shocks, Monetary Policy and Stagflation. In R. Fry, C. Jones & C. Kent (eds.), *Inflation in an Era of Relative Price Shocks*. Reserve Bank of Australia, (pp. 60–84).

Kilian, L. (2010b). Oil Price Volatility: Origins and Effects. *World Trade Organisation Staff Working Paper*, RSD-2010-02.

Kilian, L. & Lee, T.K. (2014). Quantifying the Speculative Component in the Real Price of Oil: The Role of Global Oil Inventories. *Journal of International Money and Finance, 42*, 71–87.

Kilian, L. & Murphy, D. (2014). The Role of Inventories and Speculative Trading in the Global Market for Crude Oil. *Journal of Applied Econometrics, 29*(3), 454–478.

King, H. (2014). *Trading gold and silver futures contracts*. Accessed 28 October 2014 at http://www.investopedia.com/articles/options-and-futures/.

Kishan, R.P. & Opiela, T.P. (2000). Bank Size, Bank Capital, and the Bank Lending Channel. *Journal of Money, Credit and Banking, 32*(1), 121–141.

Kiyotaki, N. & Wright, R. (1989). On Money as a Medium of Exchange. *Journal of Political Economy, 97*(4), 927–954.

Kloner, D. (2001). The Commodity Futures Modernization Act of 2000. *Securities Regulation Law Journal, 29*, 286–297.

Knittel, C.R. & Pindyck, R.S. (2013). The Simple Economics of Commodity Price Speculation. *MIT Center for Energy and Environmental Policy Research Working Paper*, No. 2013-006.

Kolodziej, M., Kaufmann, R.K., Kulatilaka, N., Bicchetti, D. & Maystre, N. (2014). Crude Oil: Commodity or Financial Asset? *Energy Economics, 46*, 216–223.

Krauss, C. & Reed, S. (2015, 31 May). Prices are down, but Saudis keep oil flowing. *New York Times*. Accessed 1 December 2015 at http://www.

nytimes.com/2015/06/01/business/energy-environment/prices-are-down-but-saudis-keep-oil-flowing.html?ref=topics&_r=0.

Krichene, N. (2002). World Crude Oil and Natural Gas: A Demand and Supply Model. *Energy Economics, 24*(6), 557–576.

Krieger, S.C. (2002). Recent Trends in Monetary Policy Implementation: A View from the Desk. *Federal Reserve Bank of New York Economic Policy Review, 8*(1), 73–76.

Krishnamurthy, A. & Vissing-Jorgensen, A. (2011). The Effects of Quantitative Easing on Interest Rates: Channels and Implications for Policy. *Brookings Papers on Economic Activity, 43*(2), 215–265.

Krugman, p. (2008). Commodity prices. *New York Times.* Accessed 29 May 2015 at http://krugman.blogs.nytimes.com/2008/03/19/commodity-prices-wonkish/?_r=0.

Kuttner, K.N. & Mosser, P.C. (2002). The Monetary Transmission Mechanism: Some Answers and Further Questions. *Federal Reserve Bank of New York Economic Policy Review, 8*(1), 15–26.

Lammerding, M., Stephan, P., Trede, M. & Wilfling, B. (2013). Speculative Bubbles in Recent Oil Price Dynamics: Evidence from a Bayesian Markov-Switching State-Space Approach. *Energy Economics, 36*, 491–502.

Lavoie, M. (1984). The Endogenous Flow of Credit and the Post Keynesian Theory of Money. *Journal of Economic Issues, 18*(3), 771–797.

Lavoie, M. (2003). A Primer on Endogenous Credit-Money. In L.-P. Rochon & S. Rossi (eds.), *Modern Theories of Money: The Nature and Role of Money in Capitalist Economies.* Cheltenham, UK and Northampton, MA, USA: Edward Elgar Publishing, (pp. 506–543).

Lavoie, M. (2006a). Endogenous Money: Accommodationist. In P. Arestis & M. Sawyer (eds.), *Handbook of Alternative Monetary Economics.* Cheltenham, UK and Northampton, MA, USA: Edward Elgar Publishing, (pp. 17–34).

Lavoie, M. (2006b). A Post-Keynesian Amendment to the New Consensus on Monetary Policy. *Metroeconomica, 57*(2), 165–192.

Lavoie, M. (2010). Changes in Central Bank Procedures During the Subprime Crisis and Their Repercussions on Monetary Policy. *Levy Economics Institute of Bard College*, Working Paper No. 606.

Lavoie, M. (2014). *Post-Keynesian Economics: New Foundations.* Cheltenham, UK and Northampton, MA, USA: Edward Elgar Publishing.

Lazonick, W. (2012). The Financialization of the US Corporation: What Has Been Lost, and How It Can Be Regained. *Institute for New Economic Thinking Research Note*, No. 7.

Lee, Y.-H. & Chiou, J.-S. (2011). Oil Sensitivity and its Asymmetric Impact on the Stock Market. *Energy, 36*(1), 168–174.

Le Heron, E. & Mouakil, T. (2008). A Post-Keynesian Stock-Flow Consistent Model for Dynamic Analysis of Monetary Policy Shock on Banking Behavior. *Metroeconomica, 59*(3), 405–440.

Li, L. (2002). Macroeconomic Factors and the Correlation of Stock and Bond Returns. *Yale International Center for Finance Working Paper*, No. 02-46.

Lizardo, R.A. & Mollick, A.V. (2010). Oil Price Fluctuations and U.S. Dollar Exchange Rates. *Energy Economics, 32*(2), 399–408.

Lombardi, M.J. & Van Robays, I. (2011). Do Financial Investors Destabilize the Oil Price? *European Central Bank*, Working Paper No. 1346.

Lucas, R.E. (1972). Expectations and the Neutrality of Money. *Journal of Economic Theory, 4*, 103–124.

Ludvigson, S., Steindel, C. & Lettau, M. (2002). Monetary Policy Transmission through the Consumption-Wealth Channel. *Federal Reserve Bank of New York Economic Policy Review, 8*(1), 117–133.

Malkiel, B.G. (2003). The Efficient Market Hypothesis and Its Critics. *Journal of Economic Perspectives, 17*(1), 59–82.

Markowitz, H. (1952). Portfolio Selection. *Journal of Finance, 7*(1), 77–91.

Marx, K. (1894/2004). *Das Kapital. Kritik der politischen Ökonomie, dritter Band*. Berlin: Akademie Verlag.

Masters, M.W. (2008). Testimony Before the Committee on Homeland Security and Governmental Affairs, United States Senate, 20 May.

McGrattan, E. (1997). Comment. In B.S. Bernanke & J. Rotemberg (eds.), *NBER Macroeconomics Annual 1997*. Cambridge, MA: MIT Press, (pp. 283–289).

McLeay, M., Radia, A. & Thomas, R. (2014). Money Creation in the Modern Economy. *Bank of England Quarterly Bulletin, 54*(1), 14–25.

Meadows, D.H., Meadows, D.L., Randers, J. & Behrens III, W.W. (1972). *The Limits to Growth*. New York: Universe Books.

Meek, R.L. (1956). Some Notes on the 'Transformation Problem'. *The Economic Journal, 66*(261), 94–107.

Metcalf, G.E. (2007). Federal Tax Policy Towards Energy. *Tax Policy and the Economy, 21*, 145–184.

Meyer, G. (2014). US stripper well operators eye closures amid low oil price. *Financial Times*. Accessed 18 August 2015 at https://www.ft.com/content/f72c4c2c-8340-11e4-b017-00144feabdc0.

Miller, J.I. & Ratti, R.A. (2009). Crude Oil and Stock Markets: Stability, Instability, and Bubbles. *Energy Economics, 31*(4), 559–568.

Mills, K.G. (2015). Growth and Shared Prosperity. *Harvard Business School*, U.S. Competitiveness Project.

Minsky, H.P. (1957). Central Banking and Money Market Changes. *Quarterly Journal of Economics, 71*(2), 171–187.

Minsky, H.P. (1982). The Financial Instability Hypothesis: Capitalist Processes and the Behavior of the Economy. In C.P. Kindleberger & J.P. Laffargue (eds.), *Financial Crises: Theory, History and Policy*. Cambridge: Cambridge University Press, (pp. 13–39).

Minksy, H.P. (1994). Financial Instability Hypothesis. In P. Arestis & M. Sawyer (eds.), *The Elgar Companion to Radical Political Economy*. Cheltenham, UK and Northampton, MA, USA: Edward Elgar Publishing, (pp. 153–157).

Mishkin, F.S. (1995). Symposium on the Monetary Transmission Mechanism. *Journal of Economic Perspectives, 9*(4), 3–10.

Mishkin, F.S. (1996). The Channels of Monetary Transmission: Lessons for Monetary Policy. *National Bureau of Economic Research Working Paper*, No. 5464.

Mishkin, F.S. (2001). The Transmission Mechanism and the Role of Asset Prices in Monetary Policy. *National Bureau of Economic Research Working Paper*, No. 8617.

Mishkin, F.S. (2006). Monetary Policy Strategy: How Did We Get Here? *National Bureau of Economic Research Working Paper*, No. 12515.

Mishkin, F.S. (2007). Will Monetary Policy Become More of a Science? *National Bureau of Economic Research Working Paper*, No. 13566.

Mohanty, M. (2014). International Transmission of Monetary Policy – An Overview. *Bank for International Settlements Papers*, No. 78, 1–24.

Moore, B.J. (1988). *Horizontalists and Verticalists: The Macroeconomics of Credit Money*. Cambridge: Cambridge University Press.

Moore, B.J. (1989). Viewpoint: Why Investment Determines Saving. *Challenge, 33*(3), 55–56.

Moore, B.J. (2008). The Demise of the Keynesian Multiplier Revisited. In C. Gnos & L.-P. Rochon (eds.), *The Keynesian Multiplier*. New York: Routledge, (pp. 120–126).

Mu, X. & Ye, H. (2011). Understanding the Crude Oil Price: How Important Is the China Factor? *The Energy Journal, 32*(4), 69–92.

Murphy, R.P. (2008). Evidence that the Fed caused the housing boom. *Ludwig von Mises Institute, Mises Daily*. Accessed 7 October 2014 at https://mises.org/library/evidence-fed-caused-housing-boom.

Murray, J. & King, D. (2012). Oil's Tipping Point has Passed. *Nature, 481*, 433–435.

NASDAQ OMX Group. (2015). *NASDAQ Composite Index [NASDAQ COM]*. Accessed 14 April 2015 at http://research.stlouisfed.org/fred2/series/NASDAQCOM.

Neely, C.J. (2011). The Large-Scale Asset Purchases Had Large International Effects. *Federal Reserve Bank of St. Louis Working Paper*, No. 2010-018C.

Newbery, D.M. (2005). Why Tax Energy? Towards a More Rational Policy. *The Energy Journal, 26*(3), 1–39.

New York Mercantile Exchange [NYMEX]. (2015). *Daily energy volume and open interest.* Accessed 21 October 2015 at http://www.cmegroup.com/market-data/volume-open-interest/energy-volume.html.

Nissanke, M. (2011). Commodity Markets and Excess Volatility: Sources and Strategies to Reduce Adverse Development Impacts. *Common Fund of Commodities.*

Nissanke, M. & Kuleshov, A. (2012). An Agenda for International Action on Commodities and Development: Issues for EU Agenda beyond the MDGs. *European Report: Development,* Background Paper.

OECD. (2014). *OECD Statistics: Consumer prices (MEI).* Accessed 5 November 2014 at http://stats.oecd.org/index.aspx?queryid=26661#.

OECD. (2015). *Consumer Price Index: Total items for the United States.* FRED, Federal Reserve Bank of St. Louis. Accessed 11 April 2015 at http://research.stlouisfed.org/fred2/series/CPALTT01USM661S.

Office of Fossil Energy. (2015). *Petroleum reserves.* Accessed 1 September 2015 at https://energy.gov/fe/office-fossil-energy.

Olsen, R.A. (1998). Behavioral Finance and its Implications for Stock-Price Volatility. *Financial Analysts Journal, 54*(2), 10–18.

OPEC. (2015). *Our mission.* Accessed 14 August 2015 at http://www.opec.org/opec_web/en/about_us/23.htm.

Palley, T.I. (1991). The Endogenous Money Supply: Consensus and Disagreement. *Journal of Post Keynesian Economics, 13*(3), 397–403.

Palley, T.I. (1993). Milton Friedman and the Monetarist Counter-Revolution: A Re-Appraisal. *Eastern Economic Journal, 19*(4), 71–82.

Palley, T.I. (1996). Accommodationism versus Structuralism: Time for an Accommodation. *Journal of Post Keynesian Economics, 18*(4), 585–594.

Palley, T.I. (2003). Monetary Control in the Presence of Endogenous Money and Financial Innovation: The Case for Asset-Based Reserve. In L.-P. Rochon & S. Rossi (eds.), *Modern Theories of Money: The Nature and Role of Money in Capitalist Economies.* Cheltenham, UK and Northampton, MA, USA: Edward Elgar Publishing, (pp. 67–83).

Panzera, F.S. (2015). *Financial Instability, Central Banking, and Macroprudential Policy: How to Achieve Financial Stability.* Unpublished doctoral dissertation, University of Fribourg, Switzerland.

Parguez, A. & Seccareccia, M. (2000). The Credit Theory of Money: The Monetary Circuit Approach. In J. Smithin (ed.), *What is Money?* London: Routledge, (pp. 101–123).

Payne, J.E. (2006–07). More on the Monetary Transmission Mechanism: Mortgage Rates and the Federal Funds Rate. *Journal of Post Keynesian Economics, 29*(2), 247–257.

Pigou, A.C. (1920). *The Economics of Welfare*. London: Macmillan.

Pigou, A.C. (1951). Some Aspects of Welfare Economics. *American Economic Review, 41*(3), 287–302.

Pilkington, P. (2013). A Stock-Flow Approach to a General Theory of Pricing. *Levy Economics Institute of Bard College*, Working Paper No. 781.

Platts. (2015). *Oil prices, industry news and analysis*. Accessed 26 November 2015 at http://www.platts.com/commodity/oil.

Pollin, R. (1991). Two Theories of Money Supply Endogeneity: Some Empirical Evidence. *Journal of Post Keynesian Economics, 13*(3), 366–396.

Pollin, R. (2012). The Great US Liquidity Trap of 2009–2011: Are We Stuck Pushing On Strings? *Review of Keynesian Economics, 0*(1), 55–76.

Pozsar, Z., Adrian, T., Ashcraft, A. & Boesky, H. (2013). Shadow Banking. *Federal Reserve Bank of New York Economic Policy Review, 19*(2), 1–16.

Ramos-Francia, M. & García-Verdú, S. (2014). The Transmission of US Monetary Policy Shocks to EMEs: An Empirical Analysis. *Bank for International Settlements Papers*, No 78, 363–397.

Reed, S. (2014, 27 November). OPEC holds production unchanged; prices fall. *New York Times*. Accessed 4 March 2015 at http://www.nytimes.com/2014/11/28/business/international/opec-leaves-oil-production-quotas-unchanged-and-prices-fall-further.html.

Reed, S. (2015, 5 June). OPEC, keeping quotas intact, adjusts to oil's new normal. *New York Times*. Accessed 1 December 2015 at http://www.nytimes.com/2015/06/06/business/international/opec-oil-prices.html?ref=topics.

Ricardo, D. (1821/1923). Grundsätze der Volkswirtschaft und Besteuerung. (3. Aufl.). In H. Waentig (Hrsg.), *Sammlung sozialwissenschaftlicher Meister*: Bd. 5. Jena: Verlag von Gustav Fischer.

Rigobon, R. & Sack, B. (2002). The Impact of Monetary Policy on Asset Prices. *National Bureau of Economic Research Working Paper*, No. 8794.

Ritter, J.A. (1995). The Transition from Barter to Fiat Money. *American Economic Review, 85*(1), 134–149.

Rochon, L.-P. (1999). The Creation and Circulation of Endogenous Money: A Circuit Dynamique Approach. *Journal of Economic Issues, 33*(1), 1–21.

Rochon, L.-P. (2004). Wicksell after the Taylor Rule: A Post-Keynesian Critique of the New Consensus. *Laurentian University Paper*.

Rochon, L.-P. (2006). Endogenous Money, Central Banks and the Banking System: Basil Moore and the Supply of Credit. In M. Setterfield (ed.), *Complexity, Endogenous Money and Macroeconomic Theory*.

Cheltenham, UK and Northampton, MA, USA: Edward Elgar Publishing, (pp. 170–186).

Rochon, L.-P. (2007). The State of Post Keynesian Interest Rate Policy: Where Are We and Where Are We Going? *Journal of Post Keynesian Economics, 30*(1), 3–11.

Rochon, L.-P. & Rossi, S. (2003). Introduction. In L.-P. Rochon & S. Rossi (eds.), *Modern Theories of Money: The Nature and Role of Money in Capitalist Economies.* Cheltenham, UK and Northampton, MA, USA: Edward Elgar Publishing, (pp. xx–lvi).

Rochon, L.-P. & Rossi, S. (2006). Inflation Targeting, Economic Performance, and Income Distribution: A Monetary Macroeconomics Analysis. *Journal of Post Keynesian Economics, 28*(4), 615–638.

Rochon, L.-P. & Rossi, S. (2011). Monetary Policy Without Reserve Requirements: Central Bank Money as Means of Final Payment on the Interbank Market. In C. Gnos & L.-P. Rochon (eds.), *Credit, Money and Macroeconomic Policy: A Post-Keynesian Approach.* Cheltenham, UK and Northampton, MA, USA: Edward Elgar Publishing, (pp. 98–115).

Rochon, L.-P. & Setterfield, M. (2011). Post-Keynesian Interest Rate Rules and Macroeconomic Performance: A Comparative Evaluation. In C. Gnos & L.-P. Rochon (eds.), *Credit, Money and Macroeconomic Policy: A Post-Keynesian Approach.* Cheltenham, UK and Northampton, MA, USA: Edward Elgar Publishing, (pp. 116–141).

Rogers, C. (2006). Exogenous Interest Rates and Modern Monetary Theory and Policy: Moore in Perspective. In M. Setterfield (ed.), *Complexity, Endogenous Money and Macroeconomic Theory.* Cheltenham, UK and Northampton, MA, USA: Edward Elgar Publishing, (pp. 290–305).

Romer, C.D. & Romer, D.H. (2004). A New Measure of Monetary Shocks: Derivation and Implications. *American Economic Review, 94*(4), 1053–1084.

Romer, D. (2000). Keynesian Macroeconomics Without the LM Curve. *Journal of Economic Perspectives, 14*(2), 149–169.

Rosa, C. (2013). The High-Frequency Response of Energy Prices to Monetary Policy: Understanding the Empirical Evidence. *Federal Reserve Bank of New York Staff Report*, No. 598.

Rossi, S. (2001). *Money and Inflation: A New Macroeconomic Analysis.* Cheltenham, UK and Northampton, MA, USA: Edward Elgar Publishing.

Rossi, S. (2003). Money and Banking in a Monetary Theory of Production. In L.-P. Rochon & S. Rossi (eds.), *Modern Theories of Money: The Nature and Role of Money in Capitalist Economies.*

Cheltenham, UK and Northampton, MA, USA: Edward Elgar Publishing, (pp. 339–359).

Rossi, S. (2006a). Cross-Border Transactions and Exchange Rate Stability. In L.-P. Rochon & S. Rossi (eds.), *Monetary and Exchange Rate Systems: A Global View of Financial Crises*. Cheltenham, UK and Northampton, MA, USA: Edward Elgar Publishing, (pp. 191–209).

Rossi, S. (2006b). The Theory of Money Emissions. In P. Arestis & M. Sawyer (eds.), *Handbook of Alternative Monetary Economics*. Cheltenham, UK and Northampton, MA, USA: Edward Elgar Publishing, (pp. 121–138).

Rossi, S. (2008). *Macroéconomie monétaire. Théories et politiques*. Genève, Zurich, Bâle: Schulthess Médias Juridiques.

Rossi, S. (2009a). Inflation Targeting and Monetary Policy Governance: The Case of the European Central Bank. In C. Gnos & L.-P. Rochon (eds.), *Monetary Policy and Financial Stability: A Post-Keynesian Agenda*. Cheltenham, UK and Northampton, MA, USA: Edward Elgar Publishing, (pp. 91–113).

Rossi, S. (2009b). Monetary Circuit Theory and Money Emissions. In J.-F. Ponsot & S. Rossi (eds.), *The Political Economy of Monetary Circuits: Tradition and Change in Post-Keynesian Economics*. London: Palgrave Macmillan, (pp. 36–55).

Rossi, S. (2015). Structural Reforms in Payment Systems to Avoid Another Systemic Crisis. *Review of Keynesian Economics, 3*(2), 213–225.

Sargent, T.J. (2008). Rational Expectations. In *The Concise Encyclopedia of Economics*. Accessed 17 September 2014 at http://www.econlib.org/library/Enc/RationalExpectations.html.

Sawyer, M. (2002a). Economic Policy with Endogenous Money. In P. Arestis, M. Desai & S. Dow (eds.), Money, Macroeconomics and Keynes. London: Routledge, (pp. 35–44).

Sawyer, M. (2002b). The NAIRU, Aggregate Demand and Investment. *Metroeconomica, 53*(1), 66–94.

Sawyer, M. (2013). Endogenous Money, Circuits and Financialization. *Review of Keynesian Economics, 1*(2), 230–241.

Scheyder, E. (2015, 6 August). Sewage flow becomes Williston's oil bust indicator. *Reuters*. Accessed 18 August 2015 at http://www.reuters.com/article/us-usa-oil-sewage-idUSKCN0QB0BJ20150807.

Schinasi, G.J. (2004). Defining Financial Stability. *International Monetary Fund Working Paper*, No. WP/04/187.

Schinasi, G.J. (2005). Preserving Financial Stability. *International Monetary Fund Economic Issues*, No. 36.

Schmitt, B. (1960). *La formation du pouvoir d'achat: l'investissement de la monnaie*. Paris: Sirey.

Schularick, M. & Taylor A.M. (2009). Credit Booms Gone Bust: Monetary Policy, Leverage Cycles and Financial Crises, 1870–2008. *National Bureau of Economic Research Working Paper*, No. 15512.

Schulmeister, S. (2012). Technical Trading and Commodity Price Fluctuations. *Österreichisches Institut für Wirtschaftsforschung*, September 2012.

Schwartz, A.J. (2002). Asset Price Inflation and Monetary Policy. *National Bureau of Economic Research Working Paper*, No. 9321.

Segal, P. (2011). Oil Price Shocks and the Macroeconomy. *Oxford Review of Economic Policy, 27*(1), 169–185.

Seton, F. (1957). The 'Transformation Problem'. *Review of Economic Studies, 24*(3), 149–160.

Setterfield, M. (2006). Is Inflation Targeting Compatible with Post Keynesian Economics? *Journal of Post Keynesian Economics, 28*(4), 653–671.

Shiller, R.J. (2003). From Efficient Markets Theory to Behavioral Finance. *Journal of Economic Perspectives, 17*(1), 83–104.

Sill, K. (2007). The Macroeconomics of Oil Shocks. *Federal Reserve Bank of Philadelphia Business Review, 33*(1), 21–31.

Silvapulle, P. & Moosa, I.A. (1999). The Relationship between Spot and Futures Prices: Evidence from the Crude Oil Market. *Journal of Futures Markets, 19*(2), 175–193.

Sims, C.A. (1992). Interpreting the Macroeconomic Time Series Facts: The Effects of Monetary Policy. *European Economic Review, 36*(5), 975–1011.

Sinai, A. (2009). Should, or Can, Central Banks Target Asset Prices? A Symposium of Views. *The International Economy*, Fall, 15–16.

Smith, A. (1776/1976). *An Inquiry into the Nature and Causes of the Wealth of Nations, Volume I*. Oxford: Clarendon Press.

Smith, J.L. (2009). World Oil: Market or Mayhem? *Journal of Economic Perspectives, 23*(3), 145–164.

Smithin, J. (2013). Keynes's Theories of Money and Banking in the *Treatise* and *The General Theory*. *Review of Keynesian Economics, 1*(2), 242–256.

Stadler, G.W. (1994). Real Business Cycles. *Journal of Economic Literature, 32*(4), 1750–1783.

Staritz, C. & Küblböck, K. (2013). Re-regulation of Commodity Derivative Markets – Critical Assessment of Current Reform Proposals in the EU and the US. *Österreichische Forschungsstiftung für Internationale Entwicklung Working Paper*, No. 45.

Stein, J.C. (1998). An Adverse-Selection Model of Bank Asset and Liability

Management with Implications for the Transmission of Monetary Policy. *The RAND Journal of Economics, 29*(3), 466–486.

Stevens, R. (2014). The Strategic Petroleum Reserve and Crude Oil Prices. *University of California, Berkeley*, Working Paper.

Stiglitz, J.E. & Weiss, A. (1981). Credit Rationing in Markets with Imperfect Information. *American Economic Review, 71*(3), 393–410.

Stockhammer, E. (2012). Rising Inequality as a Root Cause of the Present Crisis. *University of Massachusetts Amherst Political Economy Research Institute Working Paper*, No. 282.

Stoll, H.R. & Whaley, R.E. (2010). Commodity Index Investing and Commodity Futures Prices. *Journal of Applied Finance, 20*, 7–46.

Summers, L. (1986). Does the Stock Market Rationally Reflect Fundamental Values? *Journal of Finance, 41*(3), 591–601.

Svensson, L.E.O. (1986). Sticky Goods Prices, Flexible Asset Prices, Monopolistic Competition, and Monetary Policy. *Review of Economic Studies, 53*(3), 385–405.

Takáts, E. & Vela, A. (2014). International Monetary Policy Transmission. *Bank for International Settlements Papers*, No 78, 51–70.

Tang, K. & Xiong, W. (2011). Index Investment and Financialization of Commodities. *National Bureau of Economic Research Working Paper*, No. 16385.

Taylor, J.B. (1993). Discretion versus Policy Rules in Practice. *Carnegie-Rochester Conference Series on Public Policy, 39*, 195–214.

Taylor, J.B. (1995). The Monetary Transmission Mechanism: An Empirical Framework. *Journal of Economic Perspectives, 9*(4), 11–26.

Terzi, A. & Verga, G. (2006). Stock-bond Correlation and the Bond Quality Ratio: Removing the Discount Factor to Generate a 'Deflated' Stock Index. *Quaderni dell'Istituto di Economia e Finanza*, No. 67.

The Economist. (2013). *Back to the futures?* Accessed 1 September 2015 at http://www.economist.com/blogs/freeexchange/2013/02/derivatives-markets-regulation .

Tobin, J. (1969). A General Equilibrium Approach to Monetary Theory. *Journal of Money, Credit, and Banking, 1*(1), 15–29.

Tobin, J. & Brainard, W. (1990). On Crotty's Critique of q-Theory. *Journal of Post Keynesian Economics, 12*(4), 543–549.

Tokic, D. (2011). Rational Destabilizing Speculation, Positive Feedback Trading, and the Oil Bubble of 2008. *Energy Policy, 39*(4), 2051–2061.

Tomaskovic-Devey, D. & Lin, K.-H. (2011). Income Dynamics, Economic Rents, and the Financialization of the U.S. Economy. *American Sociological Review, 76*(4), 538–559.

Varoufakis, Y., Halevi, J. & Theocarakis, N.J. (2011). *Modern Political Economics: Making Sense of the Post-2008 World*. New York: Routledge.

Verleger, P.K. (2003). Measuring the Economic Impact of an Oil Release from the Strategic Petroleum Reserve to Compensate for the Loss of Venezuelan Oil Production. In United States Senate, *U.S. Strategic Petroleum Reserve, 108–18,* (pp. 249–259).

Vernengo, M. (2006). Money and Inflation. In P. Arestis & M. Sawyer (eds.), *Handbook of Alternative Monetary Economics.* Cheltenham, UK and Northampton, MA, USA: Edward Elgar Publishing, (pp. 471–489).

Vickers, D. (1960). On the Economics of Break-Even. *The Accounting Review, 35*(3), 405–412.

Villar, J.A. & Joutz, F.L. (2006). The Relationship Between Crude Oil and Natural Gas Prices. *Energy Information Adminstration, Office of Oil and Gas.*

Vlasic, B. (2008). As gas costs soar, buyers flock to small cars. *New York Times.* Accessed 19 January 2016 at http://www.nytimes.com/2008/05/02/business/02auto.html.

Volkart, R. (2008). *Corporate Finance: Grundlagen von Finanzierung und Investition.* Zürich: Versus Verlag.

Weintraub, E.R. (1975). 'Uncertainty' and the Keynesian Revolution. *History of Economic Thought, 7*(4), 530–548.

White, W. (2013). Ultra Easy Monetary Policy and the Law of Unintended Consequences. *Real-World Economics Review,* 63, 19–56.

Williamson, J. (2009). Should, or Can, Central Banks Target Asset Prices? A Symposium of Views. *The International Economy,* Fall, 17–18.

Woodford, M. (2001). The Taylor Rule and Optimal Monetary Policy. *American Economic Review, 91*(2), 232–237.

Woodford, M. (2002). Financial Market Effectiveness and the Effectiveness of Monetary Policy. *Federal Reserve Bank of New York Economic Policy Review, 8*(1), 85–94.

Woodford, M. (2009). Convergence in Macroeconomics: Elements of the New Synthesis. *American Economic Journal: Macroeconomics, 1*(1), 267–279.

World Bank. (2015). *World DataBank: Global Economic Monitor (GEM).* Accessed 11 April 2015 at http://databank.worldbank.org/data/views/variableselection/selectvariables.aspx?source=global-economic-monitor-%28gem%29#s_i.

Wray, L.R. (2006). To Fix or to Float: Theoretical and Pragmatic Considerations. In L.-P. Rochon & S. Rossi (eds.), *Monetary and Exchange Rate Systems: A Global View of Financial Crises.* Cheltenham, UK and Northampton, MA, USA: Edward Elgar Publishing, (pp. 210–231).

Wu, J.C. & Xia, F.D. (2014). Measuring the Macroeconomic Impact of Monetary Policy at the Zero Lower Bound. *Working Paper,* Accessed

29 May 2015 at https://www.frbatlanta.org/cqer/research/shadow_rate. aspx.

Zhang, Y.-J., Fan, Y., Tsai, H.-T. & Wei, Y.-M. (2008). Spillover Effect of US Dollar Exchange Rate on Oil Prices. *Journal of Policy Modeling, 30*(6), 973–991.

Index